A series of student texts in

CONTEMPORARY BIOLOGY

General Editors:

Professor E. J. W. Barrington, F.R.S.
Professor Arthur J. Willis

Principles of Environmental Physics

John L. Monteith
Ph.D., F.Inst.P., F.R.S., F.R.S.E.

Professor of Environmental Physics,
University of Nottingham School of Agriculture, Sutton Bonington
Loughborough, Leics.

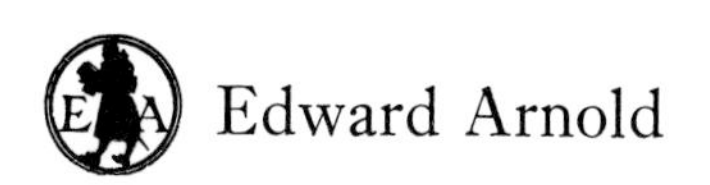
Edward Arnold

First published 1973
by Edward Arnold (Publishers) Limited
25 Hill Street
London, W1X 8LL

Boards edition ISBN: 0 7131 2374 5
Paper edition ISBN: 0 7131 2375 3

Printed in Great Britain by
William Clowes & Sons, Limited, London, Beccles and Colchester

Preface

Environmental Physics may be defined as the study of physical processes which determine how living organisms respond to their environment. When an environmental physicist is presented with a new problem, he usually begins his attack by measuring some physical component of the environment and by relating his measurement to a specific biological response. As measurements accumulate, he may attempt to formulate the general physical principles governing the behaviour of the system he is concerned with. Guided by these principles, he may then be able to forecast how his system is likely to respond to a change in its physical environment and he may suggest ways of altering the environment deliberately in order to grow more food, for example, or to make a building more comfortable. At a time when man's misuse of his environment is causing acute concern, the solution of many ecological problems must be sought in the basic principles of the physical and biological sciences and the expertise of the environmental physicist is needed now in such diverse subjects as agriculture, ecology, hydrology, medicine and building science.

Environmental research on a global scale has been stimulated by the United Nations Special Agencies and by projects such as the International Geophysical Year, the International Biological Programme, and Man and the Biosphere which have stressed the importance and demonstrated the value of communication across scientific as well as across national frontiers. Mainly as a result of international activity, the stripling subject of environmental physics is growing at a rate which has overtaken the production of textbooks reviewing the state of the art. This book was written partly to provide undergraduates in the biological sciences with a foundation for more advanced study, partly as a general source of reference for research workers, and partly to show sixth formers that the applications of classical physics to biological systems in the field can be a challenging and occasionally an entertaining pursuit.

The text is divided into three sections: a short review of relevant physics for the benefit of biologically trained readers (Chapter 2); a discussion of radiation exchange and of heat, mass and momentum transfer (Chapters 3 to 9); and the application of transfer principles to the heat balance of plants and animals and to the micrometeorology of crops (Chapters 10 to 12). Tables of quantities referred to throughout the book are provided in an appendix. Formulae have been derived from first principles wherever this was feasible. Familiarity with algebra and with the notation of elementary calculus is needed to follow these derivations but more advanced mathematics has been avoided. The more difficult parts of the book appear towards the ends of Chapters 10, 11 and 12 but it is not necessary to master these sections in order to understand the rest of the material in these chapters.

My own enthusiasm for outdoor physics was roused by James Paton in a meteorology class at Edinburgh University and was fostered by reading Rudolf Geiger's book 'The Climate Near the Ground'. Later I was fortunate to spend thirteen years in the Physics Department at Rothamsted Environmental Station where I absorbed the art and craft of environmental physics from Howard Penman. Much of this experience was formalized in a course of lectures for final year agricultural science students at the University of Nottingham and material presented in these lectures was expanded to form this book.

I am most grateful to Mr. M. F. Smith, University College of North Wales, for permission to reproduce, at the head of each chapter, short excerpts from his recent translation of a Natural Philosophy text which was published about 55 B.C.—the *De Rerum Natura* of Lucretius.

A number of authors and publishers kindly gave permission for the reproduction of diagrams. I am grateful to Mr. F. E. Lumb (Fig. 3.4), Dr. A. S. Thom (6.3, 6.4 & 12.10) and the Royal Meteorological Society; Taylor and Francis Ltd (4.7) from *Ergonomics*; Dr. E. L. Deacon and Elsevier Publishing Co. (5.1 & 8.6); Dr. S. A. Bowers and Williams & Wilkins Co. (5.3) from *Soil Science*, 1965; Dr. G. Stanhill and Pergamon Press (5.5); Dr. L. E. Mount (5.6); Deutscher Wetterdienst (5.8 & 8.4); Dr. K. Raschke and Springer-Verlag for máterial from *Planta* (7.1 & 11.6); Dr. S. Vozel and Clarendon Press (7.3); Dr. R. P. Clark and *The Lancet* (7.6 & 7.7); Dr. P. F. Scholander and the Marine Biological Laboratory (8.1); Dr. J. S. Hart and the Natural Research Council of Canada (8.2) from the Canadian Journal of Zoology; Dr. A. M. Hemmingsen (10.2 & 10.3) and Dr. E. R. Lemon (12.11); Dr. Z. Uchijima (12.12) and Pudoc, Wageningen; Dr. J. Begg and Elsevier Publishing Co. (12.13) from *Agricultural Meteorology*. Dr. A. I. Fraser and the Forestry Commission gave permission for the reproduction of Plate 3 and the photographs in Plate 4 were taken by Dr. M. H. Unsworth and Mr. A. Simms. Dr. J. V. Lake and Dr. Warren Porter provided chart records for Figs. 3.5 and 5.7 respectively and Dr. I. Impens allowed me to use unpublished measurements in Fig. 9.1.

Finally, I am glad to acknowledge help in various forms: from Miss Edna Lord who deciphered my manuscript; Dr. L. E. Mount, Dr. A. S. Thom and Dr. M. H. Unsworth, who read and suggested numerous improvements to the text; Professor A. Willis who edited the final draft with meticulous care; and the staff of Edward Arnold who handled production with patience, courtesy and skill.

J.L.M.

Sutton Bonington
1972

Table of Contents

Symbols

The main symbols used in this book have been collected in a table containing brief definitions of each quantity. A few of the symbols are universally accepted (e.g. R, g), some have been chosen because they appear very frequently in the literature of environmental physics (e.g. r_s, z_0, K_M), and some have been devised for the sake of consistency. In particular, the symbols **S** and **L** are used for flux densities of short and long wave radiation with subscripts to identify the geometrical character of the flux, e.g. $\mathbf{S}_d$ for the flux of diffuse short wave radiation from the sky.

Flux densities of momentum, heat and mass are printed in bold case throughout the book (e.g. $\boldsymbol{\tau}$, **E**) and so is the latent heat of vaporization of water $\boldsymbol{\lambda}$, partly to distinguish it from wavelength λ and partly because it is often associated with **E**. Upper case subscripts are used to refer to momentum, heat, vapour, carbon dioxide etc., e.g. r_V, K_M; all other subscripts are lower case, e.g. c_p for the specific heat of air at constant pressure.

The complete set of symbols represents the best compromise that could be found between consistency, clarity and familiarity.

Roman alphabet

A	area
A_h	area of solid object projected on a horizontal plane
A_p	area of solid object projected on plane perpendicular to solar beam
$A(z)$	amplitude of soil temperature wave at depth z

B	total energy emitted by unit area of full radiator or black body
B(λ)	energy per unit wavelength in spectrum of full radiator or black body
c	volume fraction of CO_2 (e.g. v.p.m.); fraction of sky covered by cloud; velocity of light
c_d	drag coefficient for form drag and skin friction combined
c_f	drag coefficient for form drag
c_l	specific heat of a liquid
c_p	specific heat of air at constant pressure
c_s	specific heat of solid fraction of a soil
c'	bulk specific heat of soil
C	flux of heat per unit area by convection in air
$\mathcal{C}$	heat capacity of an animal per unit surface area
d	zero plane displacement
D	diffusion coefficient for a gas in air (subscripts V, water vapour; C, CO_2); damping depth $(=(2\kappa'/\omega)^{1/2})$
e	partial pressure of water vapour in air
$e_s(T)$	saturation vapour pressure of water vapour at temperature T
δe	saturation deficit, i.e. $e_s(T)-e$
E	flux of water vapour per unit area, evaporation rate
$\mathbf{E}_o$	rate of evaporation from open water surface
$\mathbf{E}_r$	respiratory evaporation rate of animal
$\mathbf{E}_s$	rate of evaporation from skin
$\mathbf{E}_T$	rate of evaporation from vegetation
F	mass flux of a gas per unit area; flux of radiant energy
g	acceleration of gravity (9·81 m s^{-2})
G	flux of heat by conduction, per unit area
h	Planck's constant ($6{\cdot}63\times10^{-34}$ J s); relative humidity of air; height of cylinder, crop, etc.
H	total flux of sensible and latent heat, per unit area
I	intensity of radiation (flux per unit solid angle)
J	rate of change of stored heat per unit area
k	von Karman's constant (0·41); thermal conductivity of air; attenuation coefficient
k'	thermal conductivity of a solid
K	diffusion coefficient for turbulent transfer in air (subscripts H for heat, M for momentum, V for water vapour, C for CO_2)
$\mathcal{K}_s$	ratio of horizontally projected area of an object to its plane or total surface area
l	length of plate in direction of airstream
L	leaf area index; Monin–Obukhov length
L	flux of long wave radiation per unit area; (subscript u upward; d downward; e from environment; b from body)

m	weight of a molecule; air mass number
$\mathbf{M}$	rate of heat production by metabolism per unit area of body surface
M	molecular weight
M_a	molecular weight of dry air
M_v	molecular weight of water vapour
n	represents a number or dimensionless empirical constant in several equations
N	Avogadro's number ($6{\cdot}02 \times 10^{23}$); number of hours of daylight
$\mathbf{N}$	radiance (radiant flux per unit area per unit solid angle)
$\mathbf{P}$	latent heat equivalent of sweat rate per unit body area
P	total air pressure
r	radius; resistance to transfer (subscripts M momentum, H heat, V water vapour, C for CO_2); usually applied to boundary layer transfer
r_a	resistance to transfer in the atmosphere (subscripts M, H, V, C as above)
r_b	boundary layer resistance of crop for mass transfer
r_c	canopy resistance
r_d	thermal resistance of human body
r_f	thermal resistance of hair, clothing
r_h	resistance of hole (one side) for mass transfer
r_p	resistance of pore for mass transfer
r_t	total resistance of single stoma
r_H	resistance for heat transfer by convection, i.e. sensible heat
r_R	resistance for radiative heat transfer ($\rho c_p/4\sigma T^3$)
r_{HR}	resistance for simultaneous sensible and radiative heat exchange, i.e. r_H and r_R in parallel
R	Gas Constant ($8{\cdot}31$ J mol^{-1} K^{-1})
$\mathbf{R}_n$	net radiation flux density
$\mathbf{R}_{ni}$	net radiation absorbed by a surface at the temperature of the ambient air
s	relative sunfleck area
$\mathbf{S}_d$	diffuse solar irradiance on horizontal surface
$\mathbf{S}_e$	solar radiation received by a body, per unit area, as a result of reflection from the environment
$\mathbf{S}_p$	direct solar irradiance on surface perpendicular to solar beam
$\mathbf{S}_b$	direct solar irradiance on horizontal surface
$\mathbf{S}_t$	total solar irradiance (usually) on horizontal surface
t	diffusion pathlength
T	temperature
T_a	air temperature
T_b	body temperature

T_d	dew point temperature
T_e	effective temperature of ambient air
T_s, T_o	temperature of surface losing heat to environment
T_v	virtual temperature
T'	thermodynamic wet bulb temperature
u	optical pathlength of water vapour in the atmosphere
$u(z)$	velocity of air at height z above earth's surface
u^*	friction velocity
v	molecular velocity
V	velocity of a uniform airstream
$\dot{V}$	minute volume of respiratory system (ventilation rate)
w	depth of precipitable water
W	body weight of animal
x	volume fraction (subscripts s, soil; l, liquid; g, gas); ratio of cylinder height to radius;
z	distance, height above earth's surface
z_0	roughness length
Z	height of equilibrium boundary layer

Greek alphabet

α	absorption coefficient
$\alpha(\lambda)$	absorptivity at wavelength λ
β	solar elevation
γ	psychrometer constant ($=c_p p/\boldsymbol{\lambda}\varepsilon$)
γ^*	apparent value of psychrometer constant ($=\gamma r_V/r_H$)
δ	depth of a boundary layer
Δ	rate of change of saturation vapour pressure with temperature, i.e. $\partial e_s(T)/\partial T$
ε	ratio of molecular weight of water vapour and air (0·622)
ε_a	apparent emissivity of the atmosphere
$\varepsilon(\lambda)$	emissivity at wavelength λ
θ	equivalent temperature ($=T+e/\gamma$); angle with respect to solar beam
θ^*	apparent equivalent temperature ($=T+e/\gamma^*$)
κ	thermal diffusivity of still air
κ'	thermal diffusivity of a solid, e.g. soil
λ	wavelength of electromagnetic radiation
$\boldsymbol{\lambda}$	latent heat of vaporization of water
μ	coefficient of dynamic viscosity of air
$\boldsymbol{\nu}$	coefficient of kinematic viscosity of air; frequency of electromagnetic radiation
ρ	density of a gas, e.g. air including water vapour component

ρ_a	density of dry air
ρ_c	density of CO_2
ρ_l	density of a liquid
ρ_s	density of solid component of soil
ρ'	bulk density of soil
ρ	reflection coefficient
$\rho(\lambda)$	reflectivity of a surface at wavelength λ
σ	Stefan–Boltzmann constant ($5{\cdot}67 \times 10^{-8}$ W m^{-2} K^{-4})
Σ	the sum of a series
$\boldsymbol{\tau}$	flux of momentum per unit area, shearing stress
τ	fraction of incident radiation transmitted, e.g. by a leaf
ϕ	mass concentration of CO_2, e.g. g m^{-3}; angle between a plate and airstream
$\boldsymbol{\Phi}$	radiant flux density
χ	absolute humidity of air
$\chi_s(T)$	saturated absolute humidity at temperature T (°C)
ψ	angle of incidence
ω	angular frequency; solid angle

Non-dimensional Groups

Le	Lewis number (κ/D)
Gr	Grashof number
Nu	Nusselt number
Pr	Prandtl number (ν/κ)
Re	Reynolds number
Ri	Richardson number
Sc	Schmidt number ($\boldsymbol{\nu}/D$)
Sh	Sherwood number

Logarithms

ln	logarithm to the base e
log	logarithm to the base 10

Plate 1 Hoar frost on leaves of *Helleborus corsicus* (p. 8). Note the preferential formation of ice on the spikes. The exchange of heat and water vapour is faster round the edge of a leaf than in the centre of the lamina because the boundary layer is thinner at the edge (p. 80). A faster exchange of heat implies that the spikes should be somewhat warmer than the rest of the leaf, i.e. closer to air temperature at night. A faster rate of mass exchange implies that the spikes should collect hoar frost faster when their temperature is below the frost-point temperature of the air.

I

The Scope of Environmental Physics

The main obstacles are the inadequacy of our language and the novelty of my subject—factors that entail the coinage of many new terms.

LUCRETIUS, *De Rerum Natura*

To grow and reproduce successfully, organisms must come to terms with their environment. Some micro-organisms can grow at temperatures between −6 and 100°C and, when they are desiccated, can even survive down to −272°C. Higher forms of life on the other hand have adapted to a relatively narrow range of environments by evolving sensitive physiological responses to external physical stimuli. The physical environment of plants and animals has five main components which determine the survival of species:

(i) the environment is a source of radiant energy which is trapped by the process of photosynthesis in green cells and stored in the form of carbohydrates, proteins and fats. These materials are the primary source of metabolic energy for all forms of life on land and in the oceans;
(ii) the environment is a source of the water, nitrogen, minerals and trace elements needed to form the components of living cells;
(iii) factors such as temperature and daylength determine the rates at which plants grow and develop, the demand of animals for food and the onset of reproductive cycles in both plants and animals;
(iv) the environment provides stimuli, notably in the form of light or gravity, which are perceived by plants and animals and provide frames of reference both in time and in space. They are essential for resetting biological clocks, providing a sense of balance, etc;
(v) the environment determines the distribution and viability of pathogens and parasites which attack living organisms and their susceptibility to attack.

To understand and explore relationships between organisms and their environment, the Contemporary Biologist should be familiar with the main concepts of the environmental sciences. He must search for links between physiology, biochemistry and molecular biology on the one hand and meteorology, soil science and oceanography on the other. One of these links is environmental physics—the measurement and analysis of interactions between organisms and their physical environment. Interaction is the key word in this definition. The presence of an organism modifies the environment it is exposed to, so that the physical stimulus received *from* the environment is partly determined by the physiological response *to* the environment.

Several volumes would be needed to review *the* principles of environmental physics and the definite article was deliberately omitted from the title of this book because it makes no claim to be comprehensive. The topics which it covers are central to the subject, however: the exchange of radiation, heat, mass and momentum between organisms and their environment. Within these topics, similar analysis can be applied to a number of closely related problems in plant, animal and human ecology. The short bibliography at the end of the book can be consulted for more specialized treatments and for accounts of other branches of the subject such as the physics of soil water or of particulate diffusion.

The lack of a common language is often a barrier to progress in interdisciplinary subjects and it is not easy for a physicist or meteorologist with no biological training to communicate with a physiologist or ecologist who is fearful of formulae. Throughout the book, simple electrical analogues are used to describe rates of transfer and exchange between organisms and their environment, and calculus has been kept to a minimum. The concept of 'resistance' has been familiar to plant physiologists for many years, mainly as a way of expressing the physical factors that control rates of transpiration and photosynthesis, and human physiologists have used the term to describe the insulation provided by clothing or by a layer of air. In micrometeorology, aerodynamic resistances derived from turbulent transfer coefficients can be used to calculate fluxes from a knowledge of the appropriate gradients, and resistances which govern the loss of water from vegetation may soon be incorporated in models of the atmosphere that include the behaviour of the earth's surface. Ohm's Law has therefore become an important unifying principle of environmental physics, the basis of a common language for biologists and physicists. The treatment here is confined almost exclusively to steady states and to circuits in which direct current flows through resistors but there is no reason why it should not be extended to non-steady conditions by introducing appropriate alternating currents and capacities.

The choice of units was dictated by the structure of the Système

International, modified by retaining the centimetre. For example, the dimensions of leaves are quoted in cm and cm^2, diffusion resistances in $s\ cm^{-1}$ and molecular diffusion coefficients in $cm^2\ s^{-1}$. To adhere strictly to the meter or the millimetre as a unit of length often needs powers of 10 to avoid superfluous zeros and sometimes gives a false impression of precision. As most measurements in environmental physics have an accuracy between ± 1 and $\pm 10\%$, they should be quoted to 2 or at most 3 significant figures, preferably in a unit chosen to give quantities between 10^{-1} and 10^3. The area of a leaf would therefore be quoted as $23{\cdot}5\ cm^2$ rather than $2{\cdot}35 \times 10^{-3}\ m^2$ or $2350\ mm^2$. Conversions from SI to c.g.s. and British units are given in the Appendix, Table A.1.

2

Basic Physics

At no time was the fund of matter either more solidly packed or more sparse than it is now. It experiences no gains and no losses. Thus the movement of the ultimate particles now is identical to what it has been in ages past and always will be in the future.

The word 'conservation' has many shades of meaning but in the context of this book it is a basic principle of physics rather than an ecological war-cry. The physics of biological environment is based on the conservation of momentum, mass and energy in systems that contain living organisms and this introductory chapter reviews a number of important concepts, relationships and laws stemming from the principle of conservation. In the first section on Gas Laws, conservation appears in the form of Newton's Laws of Motion which are concerned with the constancy of total momentum in systems containing moving bodies. In the second section on Radiation Laws, conservation is expressed by the First Law of Thermodynamics: energy cannot be created or destroyed.

GAS LAWS

Pressure, density and temperature

Newton established the principle that when force is applied to a body, its momentum, the product of mass and velocity, changes at a rate proportional to the size of the force. Appropriately, the unit of force in the Système International is one Newton (Table A.1, p. 219). The principle of momentum conservation can be used to estimate the force exerted by gas molecules on the walls of an enclosure from the rate of change of momentum when an individual molecule strikes a wall and rebounds. By making a number of simplifying assumptions about the nature of a 'perfect' gas, it can be shown that the pressure or force per unit area of wall is

$$p = \tfrac{1}{3}\rho\overline{v^2} \qquad 2.1$$

where ρ is the density of the gas and $\overline{v^2}$ is the mean square velocity of the molecules.

Equation 2.1 resembles an empirical relation derived by combining Boyle's Law and Charles's Law:

$$p = \rho RT/M \qquad 2.2$$

where T is the absolute temperature in degrees Kelvin (K), M is the molecular weight of the gas and R is known as the Gas Constant. To emphasize the equivalence of the two equations, the molecular weight M can be expressed as the product of the number of molecules in a mole N and the weight of a single molecule m, i.e. $M = Nm$. Equation 2.1 may now be arranged in the form

$$p = \tfrac{1}{3}(N\rho/M)(m\overline{v^2}) \qquad 2.3$$

Comparison of equations 2.2 and 2.3 shows that the absolute temperature of a gas is proportional to the mean kinetic energy of individual molecules $(\tfrac{1}{2}m\overline{v^2})$ and that gas pressure is proportional (i) to molecular kinetic energy and (ii) to the number of molecules per unit volume $N\rho/M$.

In equation 2.3, p is the pressure in Newtons m^{-2} or Pascals when ρ is in g m^{-3}, M is in g, and R is in Joules mol^{-1} deg K^{-1}. A more convenient unit of pressure used in meteorology is one millibar (mbar) equal to 100 N m^{-2} and this unit is permitted in the SI.

At a standard temperature of 273 K and a standard pressure of $1{\cdot}013 \times 10^5$ N m^{-2} or 1013 mbar, the volume of one mole of any gas is

$$M/\rho = 22{\cdot}4 \times 10^{-3}\ \mathrm{m^3\ mol^{-1}} = 22{\cdot}4\ \mathrm{litres\ mol^{-1}} \qquad 2.4$$

The value of the Gas Constant is therefore

$$R = pM/T\rho = 8{\cdot}31\ \mathrm{Joules\ mol^{-1}\ deg\ K^{-1}} \qquad 2.5$$

When the molecular weight of a gas is known, its density at s.t.p. can be calculated from equation 2.4 and its density at any other temperature and pressure from equation 2.5. As a relevant example, Table 2.1 contains the molecular weights and densities at s.t.p. of the main constituents of dry air. Multiplying each density by the appropriate volume fraction gives the concentration of each component and the sum of these concentrations is the density of dry air, ρ_a. From a density of $1{\cdot}292$ kg m^{-3} and from equation 2.4, the effective molecular weight of air is 29·00 g with an error of less than 0·1%.

Table 2.1 Composition of dry air

Gas	Molecular weight g	Density at s.t.p. kg m^{-3}	Per cent by volume	Mass concentration kg m^{-3}
Nitrogen	28·01	1·250	78·09	0·975
Oxygen	32·00	1·429	20·95	0·300
Argon	38·98	1·783	0·93	0·016
Carbon dioxide	44·01	1·977	0·03	0·001
Air	29·00	1·292	100·00	1·292

Water vapour

The amount of water vapour present in air is often quoted in pressure units such as millibars or mm of mercury, but for calculating fluxes, units of concentration are more convenient. **Vapour pressure** e in mbar and concentration χ in g m^{-3} can be related by equation 2.2. Taking proper account of units and putting $M=18$, it follows that

$$\chi = \frac{10^2 eM}{RT} = \frac{217e}{T}$$

Alternatively, if the ratio of the molecular weights of water vapour and air is ε ($=0{\cdot}622$) it can be shown that

$$\chi = \rho_a \varepsilon e/(p-e) \qquad 2.6$$

where χ and ρ_a are expressed in the same units. The quantity χ is often called the **absolute humidity**.

When water is allowed to evaporate into a fixed volume of air, the concentration of water vapour increases until the air is saturated and evaporation stops. The amount of water vapour held in saturated air depends on its temperature and the symbol $\chi_s(T)$ g m^{-3} will be used for the concentration of vapour in air saturated at T°C. The corresponding vapour pressure at saturation is $e_s(T)=\chi_s(T).T/217$. Figure 2.1 shows that χ_s and e_s increase rapidly with temperature. Between 0 and 30°C, e_s increases by about $6\frac{1}{2}$% per °C whereas the pressure of an ideal gas or of unsaturated water vapour increases by only 0·4% (i.e. 1/273) per °C. The rate of increase of e_s with temperature or $\partial e_s/\partial T$ is an important quantity in micrometeorology, usually given the symbol Δ. Values of χ_s, e_s and Δ are tabulated in Table A.3 (p. 221).

The density of a mixture of air and water vapour can be found by adding the weights of air and vapour in unit volume of gas. If p is the total pressure of the mixture and e is the vapour pressure, the partial

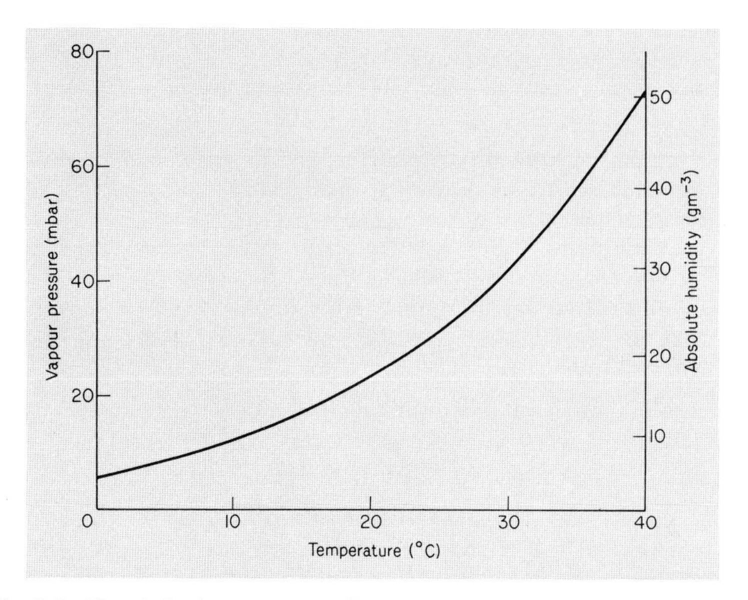

Fig. 2.1 The relation between saturation vapour pressure, absolute humidity, and temperature.

pressure of the air will be $p-e$ and its density will be $(p-e)M_a/RT$ from equation 2.3. Similarly the density of the vapour will be eM_v/RT. The density of the mixture is

$$\rho = \frac{(p-e)M_a}{RT} + \frac{eM_v}{RT}$$

$$= \frac{pM_a}{RT}\left\{1 - \frac{e(1-\varepsilon)}{p}\right\}$$

$$= \rho_a\left\{1 - \frac{e(1-\varepsilon)}{p}\right\} \qquad 2.7$$

where ρ_a is the density of dry air at the same temperature T.

Note that moist air is less dense than dry air at the same temperature. It is often convenient to express this difference in density by calculating a **virtual temperature** T_v at which dry air would have the same density as a sample of moist air at an actual temperature of T. Combining equations 2.2 and 2.7, it follows that

$$T_v = T\bigg/\left(1 - \frac{1-\varepsilon}{\varepsilon}\frac{e}{p}\right) \simeq T\left(1 + 0{\cdot}61\,\frac{e}{p}\right) \qquad 2.8$$

where $(1-\varepsilon)/\varepsilon = 0{\cdot}61$ and e is assumed to be much smaller than p. The density of dry air and the virtual temperature of saturated air between -5 and $45°$C are given in Table A.3 (p. 221).

Other parameters used to specify the amount of water vapour in air are:

(i) the **relative humidity** h which is the vapour pressure or absolute humidity divided by the corresponding value for saturated air at the same temperature, i.e. $h=\chi/\chi_s(T)=e/e_s(T)$

(ii) the **dew point** T_d which is the temperature to which unsaturated air must be cooled to produce a state of saturation, i.e. $e=e_s(T_d)$

(iii) the **saturation deficit** δe which is the difference between the saturation and actual vapour pressure of a given volume of air, i.e. $\delta e=e_s(T)-e$.

The saturation deficit, relative humidity, and dew point can be related by writing

$$\delta e = e_s(T)(1-h) \simeq \Delta(T-T_d)$$

where Δ is evaluated at a mean temperature $(T+T_d)/2$. Whereas e and χ are uniquely related to the amount of water vapour in air, relative humidity and saturation deficit depend on temperature as well as on water vapour content. Specification of water vapour content in terms of wet and dry bulb temperatures is considered in Chapter 11.

Water vapour condenses on natural surfaces which have a temperature below the dew point of the surrounding air. Dew or hoar frost is formed depending on whether the temperature of the surface is above or below the freezing point. Fog is formed when air is cooled to its dew point or when a mixture of two masses of air has a dew point higher than its temperature. When fog droplets are supercooled, they form a deposit of rime when they strike natural surfaces.

Carbon dioxide

The amount of carbon dioxide in air is usually quoted in p.p.m. (parts of CO_2 per million parts of air, by volume understood), v.p.m. (volumes per million) or μl/l (microlitres per litre), all three units being numerically identical. The relation between concentration ϕ (g m^{-3}) and volume fraction c (v.p.m.) is $\phi=\rho_c\times 10^{-6}$ where ρ_c is the density of CO_2 in g m^{-3}. At 20°C, for example, ρ_c is 1·83 kg m^{-3} and when $c=300$ v.p.m. ϕ is 0·55 g m^{-3}. The density of CO_2 at temperatures from -5 to 45°C can be found from Table A.3 (p. 221) by multiplying the density of air by $44/29=1{\cdot}52$.

The average concentration of CO_2 in the earth's atmosphere has increased steadily since accurate measurements began towards the end of the 19th century, reflecting a rapid increase in the consumption of fossil fuels. By 1964 the average annual concentration was about 314 v.p.m., and was increasing by about 0·7 v.p.m. per year. If the demand for fossil fuel continues to expand, the atmospheric concentration of CO_2 may reach 380 to 400 v.p.m. by the end of the century.[13]

Molecular transfer processes

According to equation 2.1, the mean square velocity of molecular motion in an ideal gas is $\overline{v^2}=3p/\rho$. Substituting $p=10^5$ N m^{-2} and $\rho=1{\cdot}29$ kg m^{-3} for air at 0°C gives the root mean square velocity as $(\overline{v^2})^{1/2}=480$ m s^{-1}. Molecular motion in air is therefore extremely rapid over the whole range of temperatures found in nature and this motion is responsible for a number of processes fundamental to micrometeorology: the transfer of momentum in moving air responsible for the phenomenon of viscosity; the transfer of heat by the process of conduction; and the transfer of mass by the diffusion of water vapour, carbon dioxide and other gases. Because all three forms of transfer are a direct consequence of molecular motion, they are described by similar relationships which will be considered for the simplest possible case of diffusion in one dimension only.

Viscosity

When a stream of air flows over a solid surface, its velocity increases with distance from the surface. For a simple discussion of viscosity, the velocity gradient $\partial u/\partial z$ will be assumed linear as shown in Fig. 2.2. (A more realistic velocity profile will be considered in Chapter 6.) Provided the air is isothermal, the velocity of molecular agitation will be the same

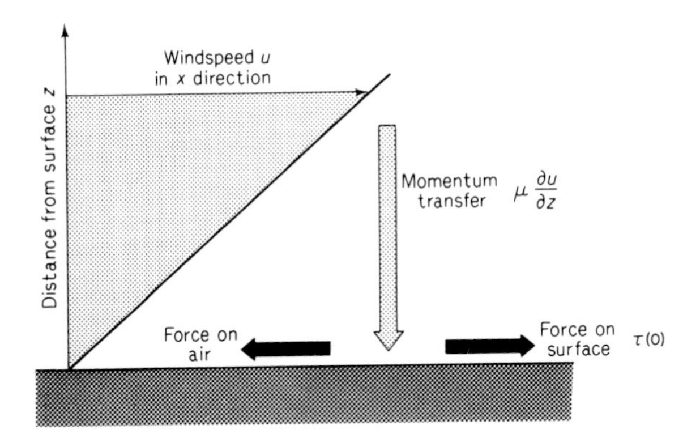

Fig. 2.2 Transfer of momentum from moving air to a stationary surface and related forces.

at all distances from the surface but the horizontal component of velocity in the x direction increases with vertical distance z. As a direct consequence of molecular agitation, there is a constant interchange of molecules between adjacent horizontal layers with a corresponding ***vertical*** exchange of *horizontal* momentum. By Newton's Second Law of Motion, a transfer of momentum produces a force proportional to the rate of transfer and, in the idealized system of Fig. 2.2, the transfer of momentum between adjacent layers of air generates a system of viscous forces proportional to the rate of change of velocity with distance, $\partial u/\partial z$. At a distance z from the surface, the viscous force per unit area $\boldsymbol{\tau}(z)$, otherwise known as the shearing stress, can be written

$$\boldsymbol{\tau}(z) = \mu\, \partial u/\partial z \tag{2.9}$$

where μ is a coefficient of dynamic viscosity (with units N s m^{-2}). At $z=0$ the surface absorbs horizontal momentum from the air, and must therefore experience a frictional force $\boldsymbol{\tau}(0)$ acting in the direction of the flow. The reaction to this force required by Newton's Third Law is the friction drag $\boldsymbol{\tau}(0)$ exerted on the air by the surface in a direction opposite to the flow.

Thermal conductivity

The conduction of heat in still air is analogous to the transfer of momentum. In Fig. 2.3, a layer of still, warm air makes contact with a cooler

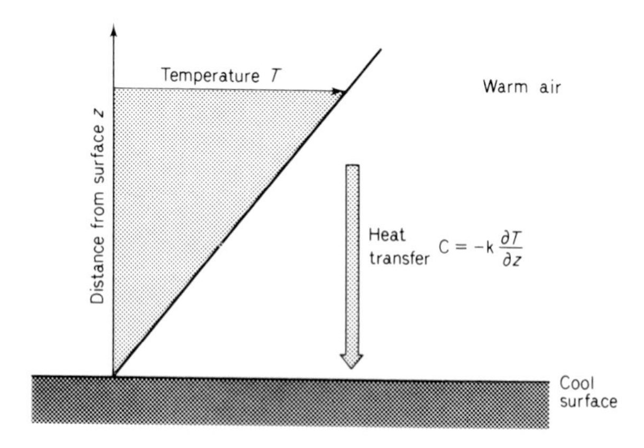

Fig. 2.3 Transfer of heat from still, warm air to a cool surface.

surface. The velocity of molecular agitation increases with distance z and the exchange of molecules between adjacent layers is responsible for an exchange of molecular energy ($\frac{1}{2}mv^2$) and hence of heat. The rate of heat transfer **C** per unit area is proportional to the temperature gradient $\partial T/\partial z$ and may therefore be written

$$\mathbf{C}(z) = -k\,\partial T/\partial z \qquad 2.10$$

where k is a coefficient of thermal conductivity with units of watts m^{-1} deg K^{-1}. The negative sign in this equation is a mathematical convention to show that when temperature increases *away from* the surface ($\partial T/\partial z > 0$), the heat flux is *towards* the surface ($\mathbf{C} < 0$).

Mass transfer

In the presence of a gradient of concentration, e.g. of water vapour or of carbon dioxide in air, molecular agitation is responsible for a transfer of mass known as diffusion. In Fig. 2.4 a layer of still air containing water

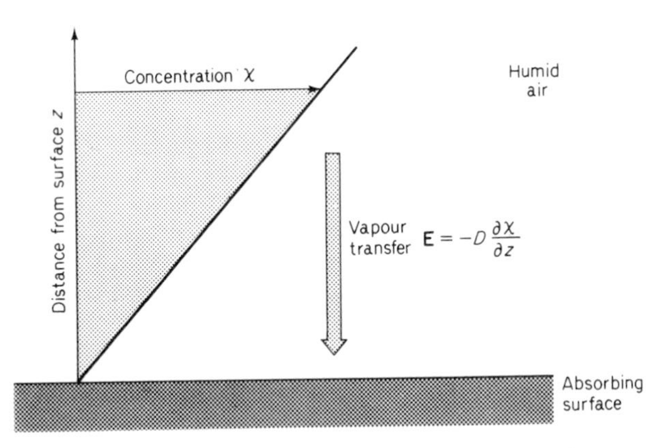

Fig. 2.4 Transfer of vapour from humid air to an absorbing surface.

vapour at a uniform temperature makes contact with a hygroscopic surface at which water vapour is absorbed. The number of molecules of water vapour per unit volume of air increases with distance from the surface and the exchange of molecules between adjacent layers produces a net movement of vapour towards the surface. The transfer of molecules expressed as a mass flux per unit area **E** is proportional to the concentration gradient $\partial \chi/\partial z$ and the transfer equation analogous to equations 2.9 and 2.10 is

$$\mathbf{E} = -D\,\partial \chi/\partial z \qquad 2.11$$

where D is a molecular diffusion coefficient for water vapour. In general, the coefficient for mass transfer depends on the molecular weights of the diffusing gas and the carrier gas and on intermolecular forces.

Diffusion coefficients

Equations 2.9, 2.10 and 2.11 express similar relationships between a flux and the gradient of a potential, but contain constants of proportionality with different dimensions; μ has units of N s m^{-2}, k is in J m^{-1} s^{-1} K^{-1}, and D is in m^2 s^{-1}. The coefficient D has the simplest dimensions because its complement $\partial\chi/\partial z$ is a concentration gradient of water vapour per unit volume of air. To reduce the equation of momentum transfer to the same form, the corresponding potential gradient is written $\partial(\rho u)/\partial z$ where ρu is the momentum per unit volume of air, i.e. a concentration of momentum. Then equation 2.9 becomes

$$\tau(z) = \frac{\mu}{\rho}\frac{\partial(\rho u)}{\partial z} = \nu\frac{\partial(\rho u)}{\partial z} \qquad 2.9a$$

where ν is the coefficient of **kinematic viscosity,** with dimensions m^2 s^{-1}.

Heat transfer can be treated in the same way by introducing the specific heat of air at constant pressure, c_p. Then ρc_p is the volumetric heat capacity of air (J m^{-3} °C^{-1}) and a change in the heat content of unit volume of air can be expressed as a change in $(\rho c_p T)$. Equation 2.10 now becomes

$$\mathbf{C} = -\frac{k}{\rho c_p}\frac{\partial(\rho c_p T)}{\partial T} = -\kappa\frac{\partial(\rho c_p T)}{\partial T} \qquad 2.9b$$

where $\kappa = k/(\rho c_p)$, called the **thermal diffusivity,** has the same dimensions as D and ν.

Because the same process of molecular agitation is responsible for all three types of transfer, the diffusion coefficients for momentum, heat, water vapour and carbon dioxide are similar in size and in their dependence on temperatures. Values of the coefficients at different temperatures calculated from the Chapman–Eskog kinetic theory of gases[8] agree well with measurements and are given in Table A.3 (p. 221). The temperature dependence of the diffusion coefficients is usually expressed by a power law, e.g. $D(T) = D(0)[T/T(0)]^n$ where $D(0)$ is the coefficient at a base temperature $T(0)$ (K) and n is an index between 1·5 and 2·0. Within the limited range of temperatures relevant to environmental physics, say -10 to 50°C, a simple temperature coefficient of 0·007 is accurate enough for practical purposes, i.e.

$$D(T)/D(0) = \kappa(T)/\kappa(0) = \nu(T)/\nu(0) = (1 + 0{\cdot}007T) \qquad 2.12$$

where T is the temperature in °C and the coefficients in units of $m^2 s^{-1}$ are

$$\nu(0) = 0{\cdot}133 \times 10^{-4} \text{ (momentum)}$$
$$\kappa(0) = 0{\cdot}181 \times 10^{-4} \text{ (heat)}$$
$$D_V(0) = 0{\cdot}212 \times 10^{-4} \text{ (water vapour)}$$
$$D_C(0) = 0{\cdot}129 \times 10^{-4} \text{ (carbon dioxide)}$$

To avoid repeating the factor 10^{-4}, the coefficients are usually quoted in units of $cm^2 s^{-1}$, e.g. $0{\cdot}13 \text{ cm}^2 \text{ s}^{-1}$.

Resistances

Equations 2.9a, 2.9b and 2.11 have the same form

$$\text{flux} = \text{diffusion coefficient} \times \text{gradient}$$

which is a general way of stating Fick's Law of Diffusion. This law can be applied to problems in which diffusion is a one-, two- or three-dimensional process but only one-dimensional cases will be considered in this book. Because the gradient of a quantity at a point is often difficult to estimate accurately, Fick's law is generally applied in an integrated form. The integration is very straightforward in cases where the (one-dimensional) flux can be treated as constant in the direction specified by the coordinate z, e.g. at right angles to a surface. Then the integration of 2.11 for example gives

$$\mathbf{E} = \frac{\chi(z_1) - \chi(z_2)}{\int_{z_1}^{z_2} dz/D} \qquad 2.13$$

where $\chi(z_1)$ and $\chi(z_2)$ are concentrations of water vapour at distances z_1 and z_2 from a surface absorbing or releasing water vapour at a rate $\mathbf{E}$. Usually $\chi(z_1)$ is taken as the concentration at the surface so that $z_1 = 0$.

Equation 2.13 and similar equations derived by integrating 2·9 and 2.10 are analogous to Ohm's Law:

$$\text{current through a resistance (amps)} = \frac{\text{potential difference across resistance (volts)}}{\text{voltage across the resistance (ohms)}}$$

Equivalent expressions for diffusion can be written as follows:

rate of transfer of		=	potential difference in units of	÷	resistance
charge	i	=	volts	÷	ohm
momentum	τ	=	ρu	÷	$\int dz/\nu$
heat	$\mathbf{C}$	=	$\rho c_p T$	÷	$\int dz/\kappa$
mass	$\mathbf{E}$	=	χ	÷	$\int dz/D$

Diffusion coefficients have dimensions of $(\text{length})^2 \times (\text{time})^{-1}$ so the corresponding resistances have dimensions of $(\text{time}) \times (\text{length})^{-1}$ or $(\text{velocity})^{-1}$. In a system where rates of diffusion are governed purely by molecular processes, the coefficients can usually be assumed independent of z so that $\int_{z_1}^{z_2} dz/D$, for example, becomes simply $(z_2 - z_1)/D$ or diffusion pathlength $\div$ diffusion coefficient. When the molecular diffusion coefficient for water vapour in air is 0·25 $\text{cm}^2\ \text{s}^{-1}$, the resistance for a pathlength of 1 cm is 4 s cm^{-1}. It is often convenient to treat the process of diffusion in laminar boundary layers in terms of resistances and in Chapters 6, 8 and 9, the following symbols will be used:

r_M resistance for momentum transfer at the surface of a body
r_H resistance for convective heat transfer
r_V resistance for water vapour transfer
r_C resistance for CO_2 transfer

The concept of resistance is not limited to molecular diffusion, however, but is applicable to any system in which fluxes are uniquely related to gradients. In the atmosphere where turbulence is the dominant mechanism of diffusion, diffusion coefficients are several orders of magnitude larger than the corresponding molecular value and increase with height above the ground. Diffusion resistances for momentum, heat, water vapour and carbon dioxide in the atmosphere will be distinguished by the symbols r_{aM}, r_{aH}, r_{aV} and r_{aC} and the measurement of these resistances is discussed in Chapters 6 to 9. In studies of the deposition of radioactive material from the atmosphere to the surface, the rate of transfer is sometimes expressed as a deposition velocity which is the reciprocal of a diffusion resistance.[21] In this case the surface concentration is often assumed to be zero and the deposition velocity is found by dividing the dosage of radioactive material by its concentration at an arbitrary height.

RADIATION LAWS

The origin and nature of radiation

Electromagnetic radiation is a form of energy derived from oscillating magnetic and electrostatic fields. It is one of the forms of energy capable of transmission through empty space where its velocity is $c = 3 \times 10^8\ \text{m s}^{-1}$ The frequency of oscillation ν is related to the wavelength λ by the standard wave equation $c = \lambda\nu$ and the wave number $1/\lambda = \nu/c$ is sometimes used as an index of frequency.

The ability to emit and absorb radiation is an intrinsic property of solids, liquids and gases and is always associated with changes in the energy state of atoms and molecules. Changes in the energy state of atomic electrons are associated with *line* spectra confined to a specific frequency or set of frequencies. In molecules, the energy of radiation is derived from the vibration and rotation of individual atoms within the molecular structure and numerous possible energy states allow radiation to be emitted or absorbed over a wide range of frequencies to form *band* spectra. The principle of energy conservation is fundamental to the material origin of radiation. The amount of radiant energy emitted by an individual atom or molecule is equal to the decrease in the potential energy of its constituents.

Reconciling the wave-like behaviour of radiation with the discrete amounts of energy absorbed and emitted by matter is one of the main achievements of modern physics. In 1900, Planck showed that radiation was emitted in quantities or 'quanta' proportional to frequency ν. The energy of a single quantum is an indivisible amount of radiation $h\nu$ and h, known as Planck's constant, has a value of $6{\cdot}63 \times 10^{-34}$ Joule s. For example, at a wavelength of 0·550 μm corresponding to the middle of the visible spectrum, $\nu = 5{\cdot}5 \times 10^{14}\ \mathrm{s}^{-1}$ and the amount of energy per quantum is $3{\cdot}6 \times 10^{-19}$ J. The amount of energy in a single quantum is inconveniently small for experimental work and a larger working unit called the Einstein is often used in photochemistry. One Einstein is N quanta where $N = 6{\cdot}02 \times 10^{23}$ is Avogadro's number, the number of molecules in a gram molecule. So a photochemical reaction needing one quantum per molecule needs one Einstein per gram molecule.

Full or black body radiation

Relationships between radiation absorbed and emitted by matter were examined by Kirchoff. He defined the *absorptivity* of a surface $\alpha(\lambda)$ as the fraction of incident radiation absorbed at a specific wavelength λ and the *emissivity* $\varepsilon(\lambda)$ as the ratio of the actual radiation emitted at λ to a hypothetical amount of radiant flux $\mathbf{B}(\lambda)$. By considering the thermal equilibrium of an object inside an enclosure at a uniform temperature, he showed that $\alpha(\lambda)$ is always equal to $\varepsilon(\lambda)$. For an object completely absorbing radiation at wavelength λ, $\alpha(\lambda) = 1$, $\varepsilon(\lambda) = 1$ and the emitted radiation is $\mathbf{B}(\lambda)$. In the special case of an object with $\varepsilon = 1$ at *all* wavelengths, the spectrum of emitted radiation is known as the 'full' or **black body** spectrum. Within the range of temperatures prevailing at the earth's surface, nearly all the radiation emitted by full radiators is confined to the waveband 3 to 100 μm, and most natural objects—soil, vegetation, water—behave radiatively like full radiators in this restricted region of the spectrum but not in the visible spectrum. Even fresh snow, one of the whitest surfaces in nature, emits radiation like a black body between 3 and 100 μm. The statement 'snow behaves like a black body' refers therefore to the radiation *emitted* by a snow surface and not to solar radiation *reflected* by snow. The semantic confusion inherent in the term 'black body' can be avoided by referring to 'full radiation' and to a 'full radiator'.

Wien's Law

After Kirchoff's work was published in 1859, the emission of radiation

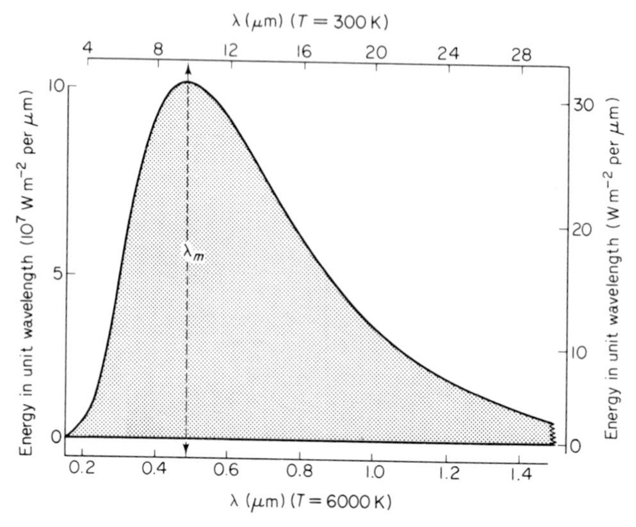

Fig. 2.5 Spectral distribution of radiant energy from a full radiator at a temperature of (a) 6000 K, left-hand vertical and lower horizontal axis and (b) 300 K, right-hand vertical and upper horizontal axis. About 10% of the energy is emitted at wavelengths longer than those shown in the diagram. If this tail were included, the total area under the curve would be proportional to σT^4 W m^{-2}. λ_m is the wavelength at which the energy per unit wavelength is maximal.

by matter was investigated by a number of experimental and theoretical physicists. The amount of energy in a narrow waveband was measured by combining a spectrometer with a sensitive thermopile, establishing that the spectral distribution of radiation from a full radiator resembles the curve in Fig. 2.5 in which the chosen temperatures of 6000 and 300 K correspond approximately to the black body temperatures of the sun and the earth's surface.

Wien deduced that the maximum energy per unit wavelength should be emitted at a wavelength λ_m given by

$$\lambda_m = 2897/T \quad \mu m \qquad 2.14$$

In Fig. 2.5, λ_m is 0·48 μm for $T=6000$ K and 9·7 μm for $T=300$ K where T is in K. A complete explanation for the distribution of energy in full radiation evaded physicists for many years but was finally provided by Planck on the basis of his quantum hypothesis.

Stefan's law

Stefan and Boltzmann showed that the energy emitted by a full radiator was proportional to the fourth power of its absolute temperature: in symbols

$$\mathbf{B} = \sigma T^4 \qquad 2.15$$

where **B** (W m^{-2}) is the flux emitted by unit area of a plane surface into an imaginary hemisphere surrounding it and the **Stefan–Boltzmann constant** σ is $5{\cdot}57 \times 10^{-8}$ W m^{-2} K^{-4}. The spatial distribution of this flux is considered on p. 19.

As a generalization from equation 2.15, the radiation emitted from unit area of any plane surface with an emissivity of ε (<1) can be written in the form

$$\mathbf{\Phi} = \varepsilon\sigma T^n$$

where n is a numerical index. For a 'grey' surface whose emissivity is almost independent of wavelength, $n=4$. When radiation is emitted predominantly at wavelengths less than λ_m, n exceeds 4 and conversely when the bulk of emitted radiation appears in a waveband above λ_m, n is less than 4.

Spatial relationships

An amount of radiant energy emitted, transmitted, or received per unit time is known as a **radiant flux** and in most problems of environmental physics, the watt is a convenient unit of flux. The term **radiant flux density** means flux per unit area, usually quoted in watts per m^2. **Irradiance** is the radiant flux density incident on a surface and **emittance** (or radiant excitance) is the radiant flux density emitted by a surface.

For a beam of parallel radiation, flux density is defined in terms of a plane at right angles to the beam, but several additional terms are needed to describe the spatial relationships of radiation dispersing in all directions from a point source or from a radiating surface. Figure 2.6(a) represents the flux **dF** emitted from a point source into a solid angle $d\omega$ where **dF** and $d\omega$ are both very small quantities. The intensity of radiation **I** is defined as the flux per unit solid angle or $\mathbf{I}=\mathbf{dF}/d\omega$. This quantity may be expressed in watts per steradian.

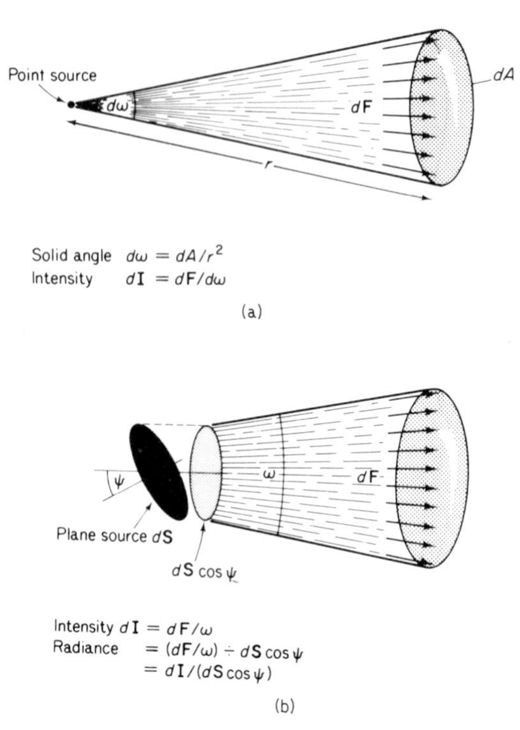

Fig. 2.6 (**a**) Geometry of radiation emitted by a point source. (**b**) Geometry of radiation emitted by a surface element. In both diagrams, a portion of a spherical surface receives radiation at normal incidence, but when the distance between the source and the receiving surface is large, it can be treated as a plane.

Figure 2.6(b) illustrates the definition of a closely related quantity—radiance. An element of surface with an area dS emits a flux **dF** in a direction specified by an angle ψ with respect to the normal. When the element is projected at right angles to the direction of the flux it shrinks to an area $dS \cos \psi$ which is the apparent or effective area of the surface viewed from an angle ψ. The **radiance** of the element in this direction is the flux emitted in the direction per unit solid angle, or $\mathbf{dF}/\omega$, divided by the projected area $dS \cos \psi$. In other words, radiance is equivalent to the intensity of radiant flux observed in a particular direction divided by the apparent area of the source in the same direction. This quantity may be expressed in watts per m^2 per steradian.

The term 'intensity' is often used colloquially as a synonym for flux density and is confused with radiance in several standard text-books.

Cosine law for emission and absorption

The concept of radiance is linked to an important law describing the spatial distribution of radiation emitted by a full radiator which has a uniform surface temperature T. This temperature determines the total flux of energy emitted by the surface (σT^4) and can be estimated by measuring the radiance of the surface with an appropriate instrument.

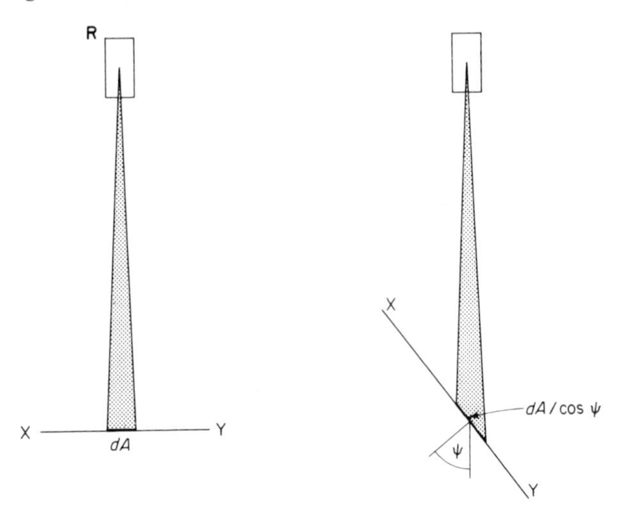

Fig. 2.7 The amount of radiation received by the radiometer from the surface XY is independent of the angle of emission, but the flux emitted per unit area is proportional to cos ψ.

As the surface of a full radiator will appear to have the same temperature whatever angle it is viewed from, the intensity of radiation emitted from a point on the surface and the radiance of an element of surface must both be independent of ψ. On the other hand, the flux per unit solid angle divided by the *true* area of the surface must be proportional to cos ψ.

Figure 2.7 makes this point diagrammatically. A radiometer R mounted vertically above an extended horizontal surface XY 'sees' an area dA and measures a flux which is proportional to dA. When the surface is tilted through an angle ψ, the radiometer now sees a larger surface $dA/\cos\psi$, but provided the temperature of the surface stays the same, its radiance will be constant and the flux recorded by the radiometer will also be constant. It follows that the flux emitted per unit area (the emittance of the surface) at an angle ψ must be proportional to $\cos\psi$ so that the product of emittance ($\propto \cos\psi$) and the area emitting to the instrument ($\propto 1/\cos\psi$) stays the same for all values of ψ.

This argument is the basis of **Lambert's Cosine Law** which states that when radiation is emitted by a full radiator at an angle ψ to the normal, the flux per unit solid angle emitted by unit surface is proportional to $\cos\psi$. As a corollary to Lambert's Law, it can be shown by simple geometry that when a full radiator is exposed to a beam of radiant energy at an angle ψ to the normal, the flux density of the absorbed radiation is proportional to $\cos\psi$.

Reflection

The reflectivity of a surface $\rho(\lambda)$ is defined as the ratio of the incident flux to the reflected flux at the same wavelength λ. Two extreme types of behaviour can be distinguished. For surfaces exhibiting specular or mirror-like reflection, a beam of radiation incident at an angle ψ to the normal is reflected at the same angle $(-\psi)$. On the other hand, the radiation scattered by a perfectly diffuse reflector is distributed in all directions according to the Cosine Law, i.e. the intensity of the scattered radiation is independent of the angle of reflection but the flux reflected from a specific area is proportional to $\cos\psi$.

The nature of reflection from the surface of an object depends in a complex way on its electrical properties and on the structure of the surface. In general, specular reflection assumes increasing importance as the angle of incidence increases and surfaces which are acting as specular reflectors absorb less radiation than diffuse reflectors made of the same material.

Most natural surfaces act as diffuse reflectors when ψ is less than 60 or 70°, but as ψ approaches 90°, a condition known as grazing incidence, the reflection from open water, waxy leaves and other smooth surfaces becomes dominantly specular and there is a corresponding increase in reflectivity. The effect is often visible at sunrise and sunset over an extensive water surface, or a lawn, or a field of barley in ear.

Radiance and irradiance

When a plane surface is surrounded by a uniform source of radiant energy, a simple relation exists between the irradiance of the surface (the flux incident per unit area) and the radiance of the source. Figure 2.8 represents a surface of unit area shown as a grey disc which is surrounded by a radiating hemispherical shell so large that the disc can be treated as a point at the centre of the hemisphere. The area dS (black) is a small element of radiating surface and the radiation reaching the centre of the hemisphere from dS makes an angle β with the normal to the plane. As the area of the grey disc projected in the direction of the radiation is $1 \times \cos \beta$, the solid angle which the disc subtends at dS is $\omega = \cos \beta / r^2$. If the element dS has a radiance $\mathbf{N}$, the flux emitted by dS in the direction of the disc must be $\mathbf{N} \times dS \times \omega = \mathbf{N}\, dS \cos \beta / r^2$. To find the total irradiance of the disc, this quantity must be integrated over the whole hemisphere, but if the radiance is uniform, conventional calculus can be avoided by noting that $dS \cos \beta$ is the area dS projected on the equatorial plane. It follows that $\int \cos \beta\, dS$ is the area of the whole plane or πr^2 so that the total irradiance at the centre of the plane becomes

$$(\mathbf{N}/r^2) \int \cos \psi\, dS = \pi \mathbf{N} \qquad 2.16$$

The irradiance expressed in watts m^{-2} is therefore found by multiplying the radiance in watts m^{-2} steradian^{-1} by the solid angle π.

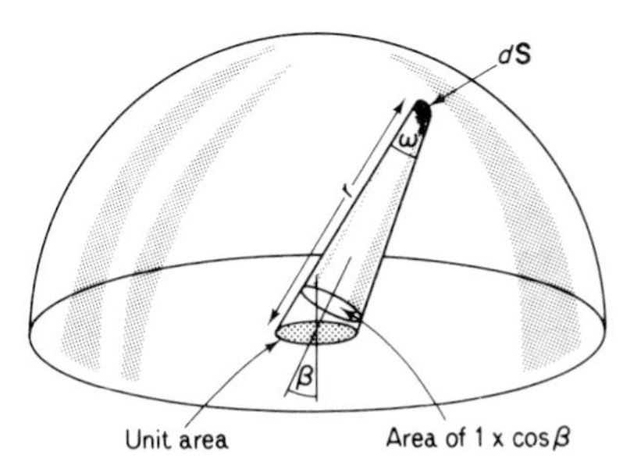

Fig. 2.8 Irradiance at the centre of an equatorial plane from the element dS of a hemispherical surface with radius r. The grey disc (unit area) receives radiation at an angle β.

A similar relation exists between the total flux of energy emitted by unit surface area of a full radiator into a hemisphere or $\mathbf{B} = \sigma T^4$ and the radiance of the surface $\mathbf{N}$. In this case $\mathbf{B} = \pi \mathbf{N}$. In several standard treatments of the subject, $\mathbf{B}/\pi$ is incorrectly described as the *intensity* at normal incidence. In fact, it is the *radiance* at normal incidence, and for a full radiator, this is the same as the radiance at any other angle.

Attenuation of a parallel beam

Beer's Law describes the transmission of a parallel beam of monochromatic radiation through a homogeneous medium. At any depth x in a medium the flux density of radiation can be written $\mathbf{\Phi}(x)$ (Fig. 2.9). The loss or attenuation of radiation in a thin layer dx is assumed proportional both to dx and to $\Phi(x)$, i.e. $d\Phi = -k\Phi(x)\,dx$ where the constant of proportionality k is an attenuation coefficient. Integration gives

$$\Phi(x) = \Phi(0)\,e^{-kx} \qquad 2.17$$

where $\Phi(0)$ is the flux incident at $x=0$.

The law can be extended to radiation in any waveband for which k is constant and is often used empirically to describe the process of attenua-

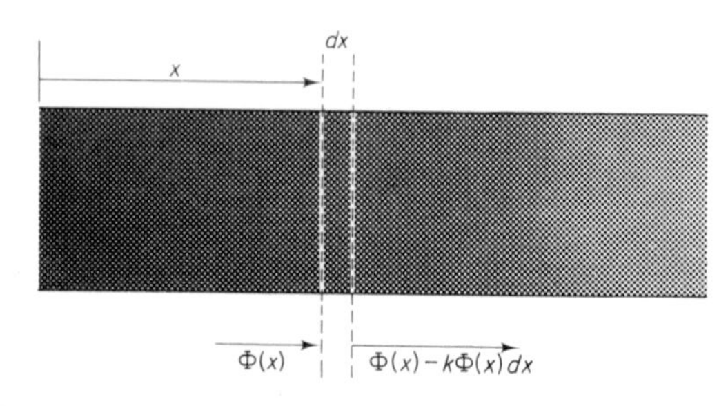

Fig. 2.9 The absorption of a parallel beam of monochromatic radiation in a homogeneous medium with an absorption coefficient k. $\Phi(0)$ is the incident flux, $\Phi(x)$ is the flux at depth x and the flux absorbed in a thin layer dx is $k\Phi(x)\,dx$.

tion in a system where k is *nearly* independent of wavelength. When k changes rapidly with wavelength, Beer's Law is invalid, but attenuation can sometimes be described accurately enough by writing the effective pathlength in the medium as x^n where n is an index less than unity. For example, the transmission of solar radiation in the earth's atmosphere can be expressed by an equation of the form

$$\Phi(x) = \Phi(0)\,e^{k_1 x^{n_1} + k_2 x^{n_2} + \cdots} \qquad 2.18$$

where x is the geometrical pathlength of the solar beam and terms of the form e^{kx^n} account for absorption and scattering by gas molecules and by aerosol.

3

The Radiation Environment

It need occasion no surprise that so small a sun can emit so great a light, sufficient to inundate all seas and lands and sky, and deluge the whole world with blazing heat.

SOLAR RADIATION

Analysis of the solar spectrum shows that the sun behaves like a full radiator with a surface temperature of about 6000 K. Most of the radiation emitted at this temperature is confined to the waveband from 0·3 to 3 μm and the amount of energy per unit wavelength is greatest at about 0·48 μm (i.e. 2897/6000, p. 17). Integrated over the whole spectrum, the energy emitted by the sun is about 74 million W m^{-2} (from σT^4, p. 17). At the mean distance of the earth from the sun, $1{\cdot}5 \times 10^8$ km, the irradiance of a surface at right angles to the solar beam is known as the **Solar Constant**, but this quantity is difficult to determine accurately from measurements below the earth's atmosphere. During the last 40 years, determinations of the Solar Constant have ranged from 1360 to 1400 W m^{-2} (1·94 to 2·00 cal cm^{-2} min^{-1}) and the latest observations[60] from balloons, from conventional jet aircraft, and from rocket aircraft flying above the stratosphere, support a value of 1360 W m^{-2}.

Spectral distribution

The solar spectrum can be divided into three main regions: the ultraviolet from 0·3 to 0·4 μm; the visible spectrum from 0·4 (blue light) to 0·7 μm (red light); and the infra-red from 0·7 to 3 μm.

Outside the atmosphere, the ultra-violet waveband contains about 7% of the energy in the solar spectrum. Quanta of radiation in the ultra-violet have a relatively large energy content (proportional to $1/\lambda$) and are responsible for several important biological effects: the deactivation of viruses, the formation of vitamin D, sunburn, and the stimulation of skin cancers. The potentially harmful effects of ultra-violet radiation are minimized by absorption within a layer of ozone in the stratosphere and by scattering in the atmosphere, processes which reduce the amount of energy in this waveband to about 3% of solar radiation at sea level.

The waveband from 0·4 to 0·7 μm contains energy, often referred to as 'light', which plays a fundamental role in biological processes such as photosynthesis and photomorphogenesis in plants, vision in animals and man, and photoperiodic effects both in the plant and in the animal kingdom. In this waveband, the main effect of the atmosphere is to scatter radiation intercepted by individual molecules of nitrogen, oxygen and other gases. Because these molecules are much smaller than the wavelength of the radiation, the scattering process obeys Rayleigh's Law: the efficiency of scattering is proportional to $1/\lambda^4$. This means that blue light at 0·4 μm is scattered about 9 times more effectively than red light at 0·7 μm. Rayleigh scattering is responsible for the blue colour of the sky seen from the earth and of the earth's atmosphere seen by astronauts on lunar expeditions.

On a cloudless day, blue skylight usually accounts for 15 to 25% of the radiation received at the earth's surface and the remaining 85 to 75% is contained in the direct solar beam—which is deficient in blue light compared with the extraterrestrial solar spectrum. Figure 3.1 shows a calculated spectral distribution for sunlight and skylight.

In the *visible* region of the spectrum, absorption by atmospheric gases is much less important than scattering in determining the spectral distribution of solar energy. In the *infra-red* spectrum, however, absorption is much more important than scattering and several atmospheric constituents absorb strongly, notably water vapour with absorption bands between 0·9 and 3 μm. The presence of water vapour in the atmosphere increases the amount of visible radiation relative to infra-red radiation.

Because the energy of solar radiation is often measured with instruments sensitive over the whole waveband from 0·4 to 3 μm, biologists need to know what fraction of this energy is received in the visible spectrum. The fraction is often assumed to be 45%, a figure derived by Moon[94] from the spectral distribution of solar radiation when the sun is more than 30° above the horizon. Moon's figures refer to direct radiation only and as diffuse sky radiation is predominantly within the visible spectrum (Fig. 3.1), the visible content of direct plus diffuse radiation should exceed 45%. When the direct and diffuse components are combined in proportions derived from Fig. 3.2, the visible fraction of the total beam is 50%. The partitioning of solar radiation between visible and infra-red

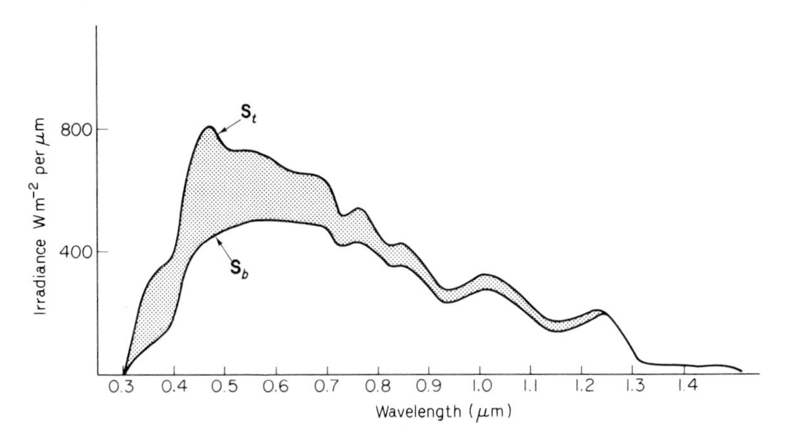

Fig. 3.1 Spectral distribution of total solar radiation (upper curve) and direct solar radiation (lower curve) calculated for a model atmosphere by Tooming and Gulyaev.[141] Solar elevation is 30° ($m = 2$) and precipitable water is 21 mm. The shaded area represents the diffuse flux which has maximum energy per unit wavelength at about 0·46 μm.

is almost independent of solar elevation because the visible fraction of skylight increases and the visible fraction of direct radiation decreases as the sun approaches the horizon.[89] For practical purposes, the energy content of photosynthetically useful radiation can be taken as half the radiative energy recorded with a conventional solarimeter. Photosynthetically active radiation is often abbreviated to PAR or PHAR.

In clouds, water droplets and ice crystals absorb and scatter direct

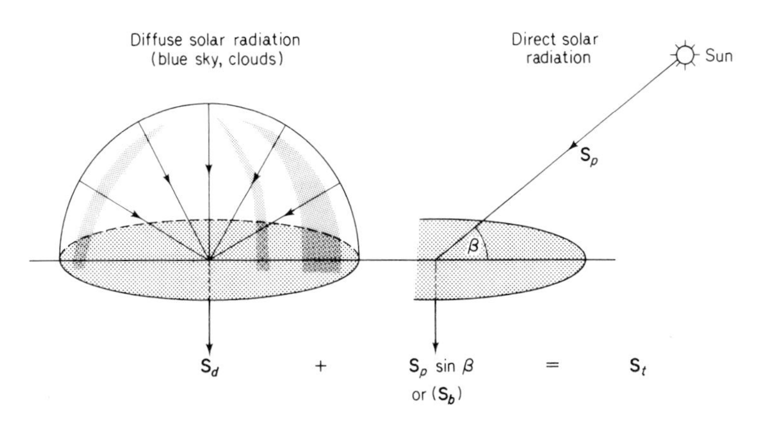

Fig. 3.2 Geometry of direct and diffuse solar radiation at the earth's surface.

solar radiation and skylight but cloud masses usually appear white or grey showing that the spectral distribution of radiation in the visible spectrum is almost unaffected by these processes. In the infra-red spectrum, liquid water absorbs radiation strongly in wavebands between 1 and 3 μm (Fig. 5.2) so the visible fraction of total solar radiation under an overcast sky tends to be slightly greater than under a cloudless sky in otherwise similar conditions. Unusual optical phenomena like irridescence or haloes produce colours in clouds by the refraction or diffraction of light but these processes involve relatively trivial amounts of energy. For dust and other particles comparable in size with the wavelength of solar radiation, the relation between scattering and wavelength is very complex. When dust accumulates in the atmosphere, within a stationary anticyclone for example, the sky becomes a much paler shade of blue, nearly white at the horizon, showing that scattering is almost independent of wavelength. Very rarely, the atmosphere contains particles of nearly uniform size that scatter red light more than blue, the converse of Rayleigh scattering. Severe forest fires in Canada in 1950 produced a pall of smoke through which both the sun and the moon looked blue. When the smoke drifted across the Atlantic, European observatories got anxious telephone calls from people who wondered if the End was Nigh!

Irradiance

Solar radiation at the earth's surface is usually measured on a horizontal plane. If $\mathbf{S}_p$ is the direct component of irradiance on a plane surface perpendicular to the solar beam at an elevation β, the direct irradiance of a horizontal surface is $\mathbf{S}_p \sin\beta$ (Fig. 3.2). (It is often more convenient to describe the position of the sun by its angle of elevation β measured with respect to the horizon instead of a zenith angle ψ. As $\beta = (\pi/2) - \psi$, $\cos\psi = \sin\beta$.) If the diffuse component of radiation from the sky and from the clouds is $\mathbf{S}_d$, the total solar irradiance can be written

$$\mathbf{S}_t = \mathbf{S}_p \sin\beta + \mathbf{S}_d = \mathbf{S}_b + \mathbf{S}_d \qquad 3.1$$

The magnitude of the components $\mathbf{S}_p$ and $\mathbf{S}_d$ depends on

(i) solar elevation,
(ii) scattering and absorption by atmospheric gases and aerosol,
(iii) scattering and absorption by water droplets and ice particles in clouds.

Elevation

The elevation of the sun at any point on the earth's surface can be calculated from its latitude, the time of day and the solar declination.[124]

The depth of atmosphere traversed by the solar beam is usually expressed by an **air mass** number m. At sea level, $m=1$ when the sun is vertical ($\beta=\pi/2$) and $m \simeq \text{cosec}\ \beta$ when $\beta > 10°$. When β is less than $10°$, m is less than cosec β by an amount determined by the curvature of the earth and by refraction. At elevated stations, m can be less than unity when the sun is near the zenith.

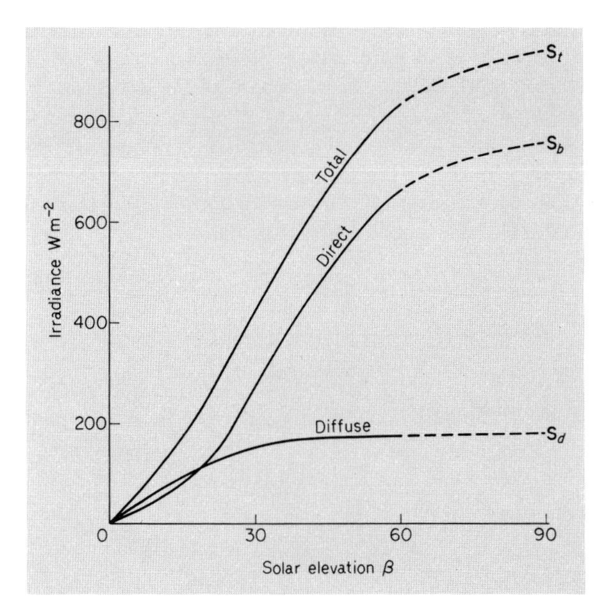

Fig. 3.3 Solar irradiance on a cloudless day (16 July 1969) at Sutton Bonington (53°N, 1°W); $\mathbf{S}_t$ total flux, $\mathbf{S}_b$ direct flux on horizontal surface; $\mathbf{S}_d$ diffuse flux. Full line from measurements; dashed line extrapolated.

To illustrate the importance of solar elevation in determining the components of irradiance, Fig. 3.3 contains smoothed measurements of $\mathbf{S}_p$ and $\mathbf{S}_b$ at Sutton Bonington (53°N, 1°W) on an unusually clear sunny day. A form of Beer's Law was used to extrapolate the measurements beyond the maximum solar elevation to 60°.

Attenuation by gases and aerosol

By extension of the concept of unit air mass, the amount of water vapour in the atmosphere can be expressed as a depth of **precipitable water** w which is the amount of rain that would be recorded at a site if all the water vapour in a vertical column of atmosphere were condensed and precipitated. When the elevation of the sun is specified by an air mass m, the equivalent pathlength for the absorption of radiation by water vapour is wm. When the pathlengths for water vapour and other gases are known, a form of Beer's Law (equation 2.17) can be used to calculate what the direct and diffuse irradiance should be when the sky is cloudless and the atmosphere is devoid of dust, smoke, and other forms of aerosol. The atttenuation by aerosol can then be calculated from *estimates* of radiation beneath a clean atmosphere and *measurements* below a real atmosphere. Even at very 'clean' stations far removed from industrial sources of pollution, the effects of aerosol are significant but in terms of the receipt of solar radiation at the surface, it is fortunate that the component of radiation scattered forwards (i.e. within 90° of the direct beam) is usually much larger than the component scattered backwards (at more than 90°). The loss of *direct* radiation by aerosol is therefore compensated by an increase of *diffuse* radiation.

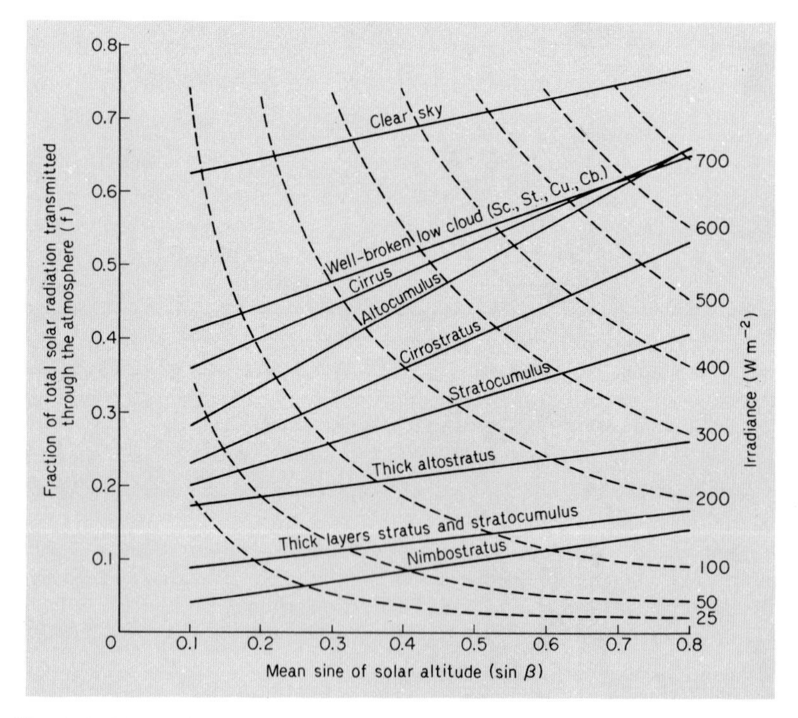

Fig. 3.4 The straight lines represent the empirical relations between solar radiation and solar angle for different cloud types from measurements in the N. Atlantic (52°N, 20°W). The curves are isopleths of irradiance (from Lumb[71]). Sc—stratocumulus, St—stratus, Cu—cumulus, Cb—cumulonimbus.

Attenuation by cloud

A deep layer of stratus cloud can reflect as much as 70% of incident sunlight from its upper surface, appearing brilliantly white like snow from an aircraft above it. About 20% of the incident radiation may be absorbed within the cloud leaving only 10% for transmission to the surface. The base of such a cloud looks grey and the transmitted radiation is wholly diffuse. At the other extreme, the total irradiance under a thin sheet of cirrus can exceed 70% of the flux under a clear sky.

Lumb[71] gave a useful summary of the effects of different cloud types on the total irradiance at different solar elevations. In Fig. 3.4 the fraction of extraterrestrial radiation can be read from the solid lines and the corresponding irradiance by interpolation from the dashed lines.

The formation of a small amount of cloud in an otherwise clear sky always increases the diffuse flux $\mathbf{S}_d$ but the direct component $\mathbf{S}_b$ remains unchanged provided the sun is not obscured by cloud. When a few isolated cumulus are present, total irradiance can therefore exceed the flux beneath a cloudless sky by 5 or 10%. As cloud amount increases, $\mathbf{S}_d$ increases to a maximum and then tends to decrease as the cover becomes complete. On a day of broken cloud, the distribution of radiation is strongly bimodal: the irradiance is very weak when the sun is completely occluded and strong when it is exposed (Fig. 3.5). For a few minutes before and after the sun is occluded the irradiance tends to be anomalously

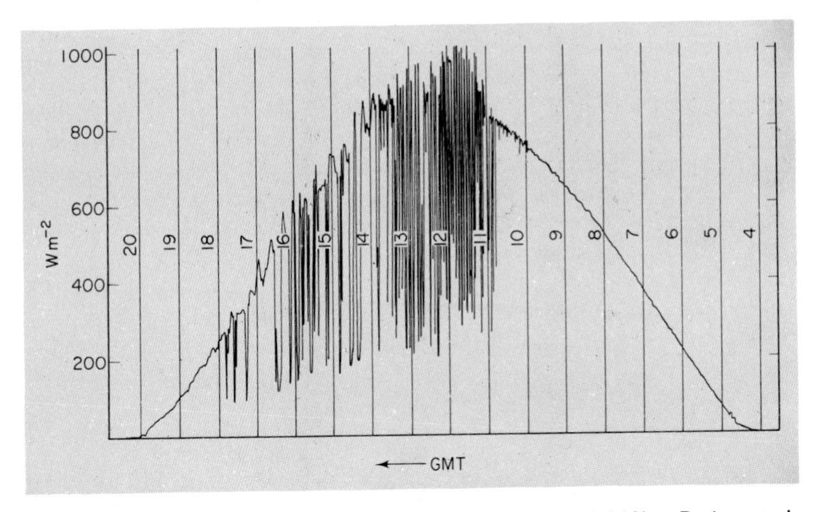

Fig. 3.5 Solar radiation on a day of broken cloud (11 June 1969) at Rothamsted (52°N, 0°W) taken directly from recorder charts. Note very high values of irradiance immediately before and after occlusion of the sun by cloud and the regular succession of minimum values when the sun is completely obscured.

large, commonly reaching 1000 W m^{-2} in temperate latitudes and sometimes even exceeding the Solar Constant in the tropics. This effect is the result of strong forward scattering of radiation by water droplets near the edge of the cloud.

The ratio of diffuse to total radiation

When the sky is cloudless, the ratio of diffuse to total radiation $\mathbf{S}_d/\mathbf{S}_t$ depends mainly on the elevation of the sun and on the amount of aerosol suspended in the atmosphere. When the sun is just below the horizon, $\mathbf{S}_d/\mathbf{S}_t$ is unity because all the radiation is diffuse. As the angle of elevation increases, $\mathbf{S}_d/\mathbf{S}_t$ decreases to a minimum value, reached when β exceeds about 40°. In very clear dry air this minimum can be as small as 0·1; it is commonly about 0·15 and in very dirty air it can reach 0·25.[34]

Angular distribution of skylight

Under a cloudless sky, the angular distribution of skylight depends on the position of the sun and cannot be described by any simple relation. In general, the sky round the sun is much brighter than elsewhere because there is a preponderance of scattering in a forward direction but there is a sector of sky about 90° from the sun where the intensity of skylight is below the average for the hemisphere. On average, the diffuse radiation from a blue sky tends to be stronger nearer the horizon than at the zenith.

Under an overcast sky, the flux of solar radiation received at the ground is almost completely diffuse. If it were perfectly diffuse, the radiance of the cloud base observed from the ground would be uniform and would therefore be equal to $\mathbf{S}_d/\pi$ from equation 2.16. The source providing this distribution is known as a Uniform Overcast Sky (UOC). In practice, the average radiance of a heavily overcast sky is about three times greater at the zenith than at the horizon. To allow for this variation, ambitious architects and meticulous microclimatologists assume that the radiance of the cloud base is proportional to $(1+2\sin\beta)/3$ and this arbitrary weighting factor defines a Standard Overcast Sky (SOC). It can be shown by integration that the radiance of the cloud at an elevation β is

$$\mathbf{N}(\beta) = 3\mathbf{S}_d(1+2\sin\beta)/7\pi \text{ W m}^{-2}\text{ steradian}^{-1} \qquad 3.2$$

where $\mathbf{S}_d$ is the total diffuse flux on a horizontal surface.

Total irradiance—instantaneous values

Total irradiance in the absence of cloud can be calculated from the behaviour of a model atmosphere or from tables, provided the relevant atmospheric parameters are known. An example of the increase of $\mathbf{S}_t$ with solar elevation is given in Fig. 3.3 derived from measurements of $\mathbf{S}_p$ and $\mathbf{S}_d$ on a day of average turbidity, (the values for $\beta > 60°$ were obtained by extrapolation using equation 2.18). Note that between 0° and 50°, $\mathbf{S}_t$ increases almost linearly with elevation.

Total irradiance—daily integrals

On cloudless days, illustrated by Fig. 3.6, the diurnal variation of total irradiance is approximately sinusoidal. In the presence of cloud, the

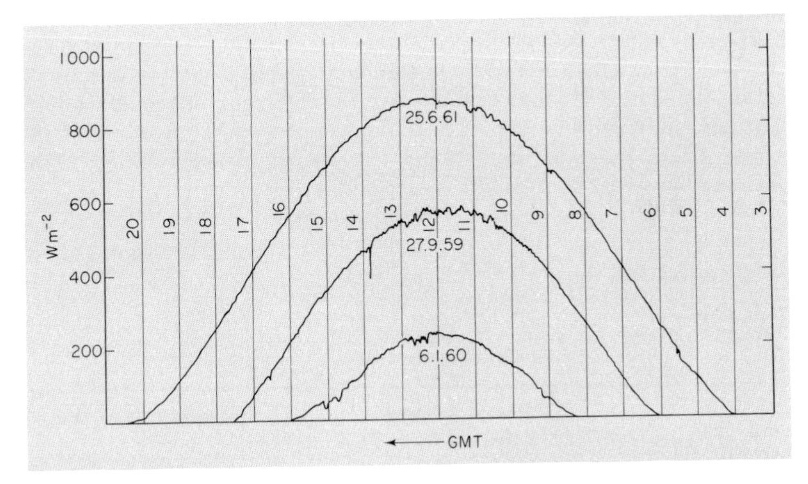

Fig. 3.6 Solar radiation on three cloudless days at Rothamsted (52°N, 0°W). During the middle of the day, the record tends to fluctuate more than in the morning and evening, suggesting a diurnal change in the amount of dust in the lower atmosphere, at least in summer and autumn. Three recorder charts were superimposed to facilitate this comparison.

radiant flux changes irregularly from minute to minute and from hour to hour. In many climates, however, the degree of cloud cover, averaged over a period of a month, is almost constant throughout the day so the diurnal change of irradiance is nearly sinusoidal when it is expressed as a monthly average.[151] The flux of total solar radiation $\mathbf{S}_t$ at t hours after sunrise can therefore be expressed as

$$\mathbf{S}_t \simeq \mathbf{S}_{tm} \sin (\pi t/N) \qquad 3.3$$

where $\mathbf{S}_{tm}$ is the maximum irradiance at solar noon and N is the day-length in hours. Equation 3.3 can be integrated to give an approximate relation between maximum irradiance and the daily integral of irradiance, conveniently referred to as daily insolation. The integral is

$$\int_0^N \mathbf{S}_t \, dt \simeq 2\mathbf{S}_{tm} \int_0^{N/2} \sin(\pi t/N) \, dt = (2N/\pi)\mathbf{S}_{tm} \qquad 3{\cdot}4$$

For example, over southern England in summer, $\mathbf{S}_{tm}$ may reach 900 W m^{-2} on a cloudless day, and $N = 16$ hours or 58×10^3 s, so the approximate insolation from equation 3.4 is 33 MJ m^{-2}. The measured maximum insolation is about 30 MJ m^{-2}. In Israel, $\mathbf{S}_{tm}$ reaches 1050 W m^{-2} in summer. For a daylength of 14 hours, equation 3.4 gives an insolation of 34 MJ m^{-2} compared with 32 MJ m^{-2} by measurement.

In many climates, the daily receipt of solar radiation is greatly reduced by cloud. Over much of Europe in summer, average insolation is between 15 and 25 MJ m^{-2}, about 50 to 80% of the insolation on cloudless days. Comparable figures in the U.S.A. range from 23 MJ m^{-2} round the Great Lakes to 31 MJ m^{-2} under the almost cloudless skies of the Sacramento and San Joaquim valleys. Winter values range from 1 to 5 MJ m^{-2} over most of Europe and from 6 MJ m^{-2} in the northern U.S.A. to 12 MJ m^{-2} in the south. Australian stations record a range of values similar to those of the U.S.A.[123, 124]

Table 3.1 Short wave radiation balance of atmosphere and surface at Kew Observatory (51·5°N) for 1956–1960 expressed as a percentage of extraterrestrial flux

	Winter (Nov–Jan)		Spring (Feb–April)		Summer (May–July)		Autumn (Aug–Oct)		Year	
Extraterrestrial radiation										
three month total MJ m^{-2}	800		2050		3720		2340		8910	
daily mean MJ m^{-2} day^{-1}	8·7		22·3		40·4		25·4		24·4	
Losses in atmosphere (per cent)										
(a) Absorption										
water vapour	15		12		13		15		13	
cloud	8		9		9		9		9	
dust and smoke	15		10		5		8		8	
total		38		31		27		32		30
(b) Scattering (away from surface)		37		35		33		34		34
Radiation at surface										
direct	8		14		18		14		15	
diffuse	17		20		22		20		21	
total		25		34		40		34		36
		100		100		100		100		100
total as MJ m^{-2} day^{-1}		2·2		7·6		16·2		8·7		8·8

At any station, the way in which insolation changes during the year depends in a complex way on seasonal changes in the water vapour and aerosol content of the atmosphere and on the seasonal distribution of cloud. Table 3.1 shows the main components of attenuation for four 'seasons' at Kew Observatory, a surburban site 10 miles (16 km) west of the centre of London.[83] For the annual average, roughly a third of the radiation received outside the atmosphere is scattered back to space, a third is absorbed and a third is transmitted to the surface. The flux at the surface is 20 to 25% less than it would be in a perfectly clear atmosphere. Because the climate is relatively cloudy, the diffuse component is larger than the direct component throughout the year.

In the absence of measurements, insolation can be estimated from cloudiness or sunshine hours and from a knowledge of the extraterrestrial irradiance or of the insolation on cloudless days.

LONG WAVE RADIATION

Most natural surfaces can be treated as 'full' radiators which emit 'terrestrial' or long wave radiation in contrast to the solar or short wave

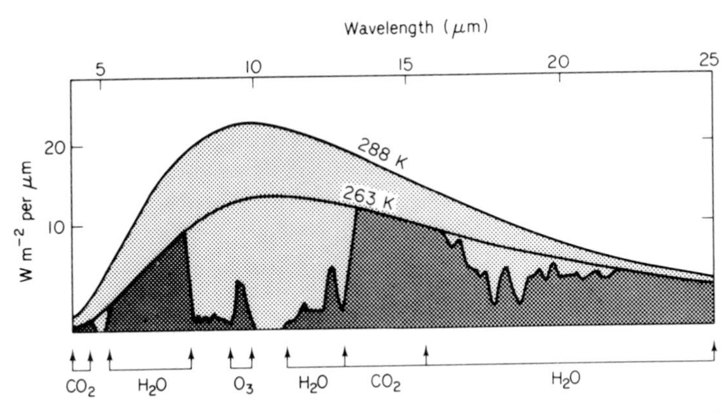

Fig. 3.7 Spectral distribution of long wave radiation for black bodies at 288 K and 263 K. Dark grey areas show the emission from atmospheric gases at 263 K. The light grey area therefore shows the net loss of radiation from a surface at 288 K to a cloudless atmosphere at a uniform temperature of 263 K (after Gates[45]).

radiation emitted by the sun. At a surface temperature of 288 K the energy per unit wavelength of terrestrial radiation reaches a maximum at 2897/288 or 10 μm (Fig. 3.7) and arbitrary limits of 3 and 100 μm are usually taken to define the long wave spectrum.

Most of the radiation emitted by the earth's surface is absorbed in specific wavebands by atmospheric gases, mainly water vapour and carbon dioxide. These gases have an emission spectrum similar to their absorption spectrum (Kirchoff's principle, p. 15) and Fig. 3.7 shows the approximate spectral distribution of the downward flux of atmospheric radiation at the earth's surface. Part of the radiation emitted by the atmosphere is lost to outer space and in order to satisfy the First Law of Thermodynamics for the earth as a planet, the average annual loss of energy must be equal to the average net gain from solar radiation.

Analysis of the exchange and transfer of long wave radiation throughout the atmosphere is one of the main problems of physical meteorology but micrometeorologists are concerned primarily with the simpler problem of measuring or estimating fluxes at the surface. The upward flux from a surface $\mathbf{L}_u$ can be measured with a radiometer or from a knowledge of the surface temperature and emissivity. The downward flux from the atmosphere $\mathbf{L}_d$ can also be measured radiometrically, or calculated from a knowledge of the temperature and water vapour distribution in the atmosphere, or estimated from empirical formulae.

Clear skies

It is often convenient to express the downward flux of atmospheric radiation as a fraction of full radiation at screen temperature (σT_a^4) and to define $\varepsilon_a = \mathbf{L}_d/\sigma T_a^4$ as the apparent emissivity of the atmosphere. When the sky is cloudless, the emissivity $\varepsilon_a(0)$ depends in principle on the distribution of temperature, water vapour and carbon dioxide in the lower atmosphere, say to a height of 5000 m. About half the radiation comes from the lowest 100 m, so ε is strongly correlated with screen temperature. Swinbank[130] showed that clear sky measurements from a number of stations fitted the relation

$$\mathbf{L}_d = 1{\cdot}20\,\sigma T_a^4 - 171$$

implying that

$$\varepsilon_a(0) = 1{\cdot}20 - 171/\sigma T_a^4 \qquad 3{\cdot}5$$

where $\mathbf{L}_d$ and σT_a^4 are in W m^{-2}. For values of T_a between 268 and 298 K, the dependence of $\varepsilon_a(0)$ on T_a is almost linear and equation 3.5 can be transformed without significant loss of accuracy to give

$$\varepsilon_a(0) = 0{\cdot}65_5 + 0{\cdot}07(T_a - 273) \qquad 3{\cdot}6$$

The factor $(T-273)$ is simply the temperature in °C but is written in this form as a reminder that T should be expressed in K when using

Stefan's Law. As a further simplification, $\mathbf{L}_d$ increases almost linearly with temperature and when T_a is expressed in °C

$$\mathbf{L}_d = 208+6T_a \text{ (W m}^{-2}\text{)} \qquad 3.7$$

in the range $-5 < T_a < 25$°C (Fig. 3.8). The error in using equation 3.6 or 3.7 instead of 3.5 is less than $\pm 1.5\%$ and is therefore comparable with

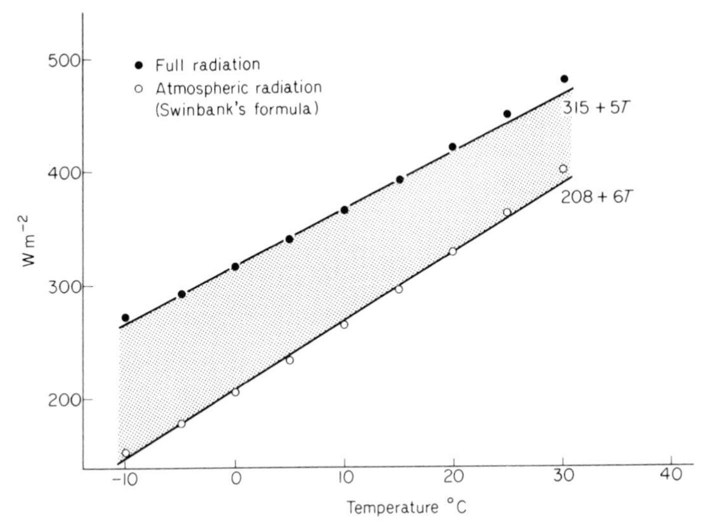

Fig. 3.8 Linear approximations to (a) σT^4 (filled circles) and (b) $\mathbf{L}_d$ from Swinbank's radiation formula (open circles). The shaded area represents the net loss of long wave radiation in the absence of cloud assuming that the screen temperature and ground temperature are both T (°C).

the standard deviation of the original measurements. Within the same temperature range and with an error of less than $\pm 1\%$, the emission from a full radiator at temperature T_b°C can be expressed as

$$\sigma T_b^4 = 315+5.0T_b \text{ (W m}^{-2}\text{)} \qquad 3.8$$

where T_b is again in °C. The effective radiative temperature of the atmosphere can now be found by equating σT_b^4 and $\mathbf{L}_d$ to give

$$T_b = \frac{208+6T_a-315}{5.0} = (T_a-21)+0.2T_a \qquad 3.9$$

This formula provides a very simple way of estimating what is sometimes loosely called 'sky temperature', and shows that when T_a is between 0 and 20°C, the sky temperature in the absence of cloud is about 20°C *below* air temperature.

The approximate net loss of radiation from a horizontal surface at air temperature is

$$\mathbf{L}_u - \mathbf{L}_d = 107 - T_a \qquad 3{\cdot}10$$

ranging from 107 to 87 W m^{-2} between 0 and 20°C. A net loss of 100 W m^{-2} is therefore a good average figure to remember for the long wave radiative loss under a cloudless sky.

The apparent emissivity of the atmosphere increases with increasing air temperature because T_a is strongly correlated with the total water content of the atmosphere. To relate ε_a to precipitable water w, it is necessary to allow for the effect of atmospheric pressure on the width of water vapour absorption bands and, because the bands get narrower as pressure decreases, the effective length of the water vapour path u, sometimes called the optical depth, is about $0{\cdot}8w$. At Kew Observatory the relation between $\varepsilon_a(0)$ and u is[81]

$$\varepsilon_a(0) = 0{\cdot}70 + 0{\cdot}22 \log u \qquad 3{\cdot}11$$

From a correlation between mean values of u and mean screen temperature in different air masses, it is possible to derive an expression similar to equation 3.5. Alternatively, it can be shown that w is well correlated with the square root of the mean vapour pressure at screen height, e (mbar), and that

$$\varepsilon_a(0) = 0{\cdot}53 + 0{\cdot}06\sqrt{e} \qquad 3{\cdot}12$$

This type of formula was first derived by Brunt.[18] Because the 'constants' in this formula depend on the distribution of temperature and water vapour in the atmosphere, different values have been derived from measurements in different climatic regions.

The intensity of long wave radiation (i.e. the flux per unit solid angle from a specific direction) increases from the zenith to the horizon because the effective pathlength for water vapour emission increases with the air mass. At an elevation β the pathlength is $um = u \operatorname{cosec} \beta$ and it has been shown[39] that the emission from a *column* of water vapour of length um is equal to the emission from a *slab* of thickness $um/1{\cdot}66$. The emissivity of a column of atmosphere at an elevation β is therefore

$$\begin{aligned}\varepsilon_a(0, \beta) &= 0{\cdot}7 + 0{\cdot}22 \log (u \operatorname{cosec} \beta/1{\cdot}66) \\ &= \varepsilon_a(0) + [0{\cdot}22 \log (u/\operatorname{cosec} \beta)] - 0{\cdot}48 \qquad 3{\cdot}13\end{aligned}$$

When $\beta = 37°$, $0{\cdot}22 \log \operatorname{cosec} \beta = 0{\cdot}48$ and $\varepsilon_a(0, \beta) = \varepsilon_a(0)$. This implies that the apparent emissivity of the atmosphere can be obtained from a single measurement of the intensity of long wave radiation at an elevation of 37°, provided the whole sky is free from cloud.[33]

Cloudy skies

Clouds dense enough to cast a shadow on the ground emit like full radiators at the temperature of the water droplets or ice crystals from which they are formed. The presence of cloud increases the flux of atmospheric radiation received at the surface because the radiation from water vapour and carbon dioxide in the lower atmosphere is supplemented by emission from clouds in the waveband which the gaseous emission lacks, i.e. from 8 to 13 μm (see Fig. 3.7).

When a fraction c of the sky is covered by cloud, the apparent emissivity $\varepsilon_a(c)$ can be expressed empirically by

$$\varepsilon_a(c) = \varepsilon_a(0)(1+nc^2) \qquad 3.14$$

where $\varepsilon_a(0)$ is calculated from equation 3.6 or 3.12.[124] The parameter n allows for the decrease of cloud temperature with increasing height: it ranges from 0·2 for low cloud (stratus, cumulus, altostratus, altocumulus) to 0·04 for cirrus type cloud. Under a heavily overcast sky, $\varepsilon_a(c)$ is often close to unity because the cloud base approaches thermal equilibrium with the surface of the ground below it.

When the frequency of different cloud types is relatively constant from month to month, a simpler linear expression can be used. At Kew, the mean radiative temperature of clouds is about 2°C less than mean screen temperature and corresponds to a flux difference of about 9 W m^{-2}.[81] The mean flux of atmospheric radiation is therefore

$$\mathbf{L}_d = (1-c)\varepsilon_a(0)\sigma T^4 + c[(\sigma T_a{}^4)-9] \text{ W m}^{-2} \qquad 3.15$$

where the first term is the flux received from clear sky and the term in square brackets is the flux from cloud. An alternative form of this equation can be derived from equation 3.8.

Limitations of empiricism

It is important to remember that the formulae presented in this section are statistical correlations of radiative fluxes with weather parameters at particular sites and do not describe direct functional relationships. For prediction, they are most accurate under average conditions, e.g. when the air temperature does not increase or decrease rapidly with height near the surface and when the air is not unusually dry or humid. They are therefore appropriate for climatological studies of radiation balance but are often not accurate enough for micrometeorological analyses over periods of a few hours. In particular, the equations cannot be used to investigate the diurnal variation of $\mathbf{L}_d$. At some sites, the amplitude of $\mathbf{L}_d$ in clear weather is much smaller than the amplitude of $\mathbf{L}_u$, behaviour to be expected if changes of atmospheric temperatures were governed by and followed changes of surface temperature. At other sites, $\mathbf{L}_d$ appears to change *more* than $\mathbf{L}_u$ for reasons which are not yet understood.

4

Radiation Geometry

Our own shadows, moreover, appear to move with us in the sunlight, following our footsteps and mimicking our gestures. . . . The fact is that particular parts of the ground are successively deprived of the light of the sun which we intercept wherever we move.

In conventional problems of micrometeorology, fluxes of radiation at the earth's surface are measured and specified by the receipt or loss of energy per unit area of a horizontal plane. To estimate the amount of radiation intercepted by the surface of a plant or animal, the horizontal irradiance must be multiplied by a shape factor depending on (i) the geometry of the surface and (ii) the directional properties of the radiation. To make the analysis more manageable, objects like spheres or cylinders with a relatively simple geometry are often used to represent the more irregular shapes of plants and animals. Appropriate shape factors for these models will be derived.

The radiation intercepted by an organism or its analogue can be expressed as the mean flux per unit area of surface. A bar will be used to distinguish this measure of irradiance from the conventional flux on a horizontal surface, e.g. when solar irradiance is $\mathbf{S}$ (W per m^2 horizontal area), the corresponding irradiance of a sheep or a cylinder will be written $\bar{\mathbf{S}}$ (W per m^2 total area).

DIRECT SOLAR RADIATION

The flux of direct solar radiation is usually measured either on a horizontal plane ($\mathbf{S}_b$) or on a plane perpendicular to the sun's rays ($\mathbf{S}_p$). For any solid object exposed to direct radiation, a simple relation between the mean flux $\bar{\mathbf{S}}_b$ and the horizontal flux $\mathbf{S}_b$ can be derived from the relation between the area of shadow A_h cast on a horizontal surface and the area projected in the direction of the beam A_p.

The area projected in the direction of the sun's rays is $A_p = A_h \sin\beta$ (Fig. 4.1) and the intercepted flux is

$$A_p\mathbf{S}_p = (A_h \sin\beta)\mathbf{S}_p = A_h\mathbf{S}_b$$

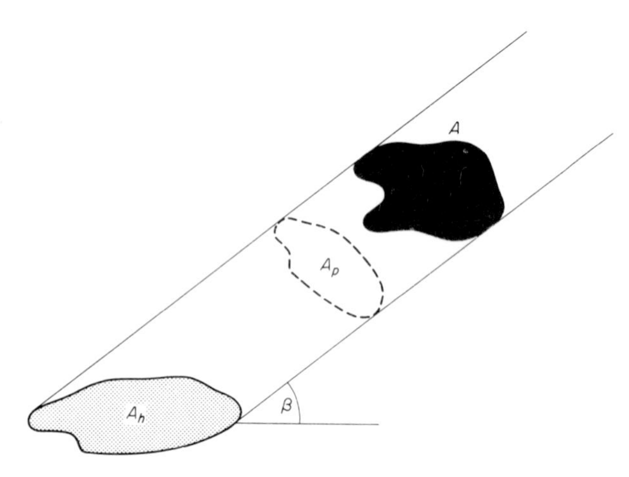

Fig. 4.1 Area A projected on surface at right angles to solar beam (A_p) and on horizontal surface (A_h).

i.e. the area of shadow on a horizontal plane times the irradiance of that plane. Then if the surface area of the object is A

$$\bar{\mathbf{S}}_b = (A_h/A)\mathbf{S}_b \qquad 4.1$$

The shape factor (A_h/A) can be calculated from geometrical principles or measured directly from the area of a shadow or from a shadow photograph.

Shape factors

Sphere

The shadow cast by a sphere of radius r has an area of $\pi r^2/\sin\beta$ (Fig. 4.2). The surface area of the sphere is $4\pi r^2$ so

$$\frac{A_h}{A} = \frac{\pi r^2}{4\pi r^2 \sin\beta} = 0{\cdot}25 \operatorname{cosec}\beta$$

The mean irradiance of a sphere is therefore

$$\bar{\mathbf{S}}_b = (0{\cdot}25 \operatorname{cosec}\beta)\mathbf{S}_b = 0{\cdot}25\mathbf{S}_p \qquad 4.2$$

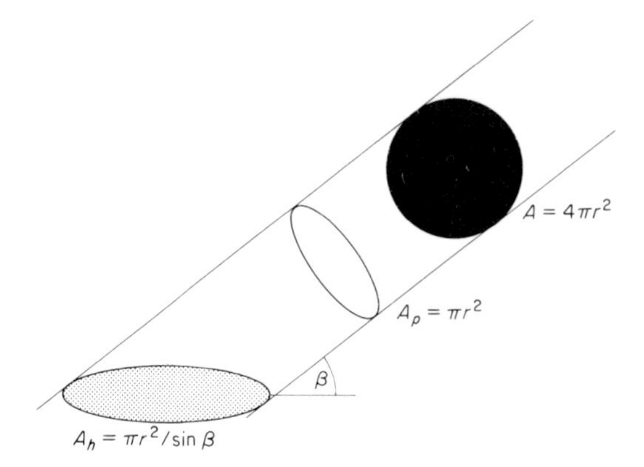

Fig. 4.2 Geometry of sphere projected on horizontal surface.

Inclined plane

Figure 4.3 shows the end and plan views of a square plane with sides of unit length making an angle α with the horizon XY and exposed to a beam of radiation at an elevation β at right angles to the edge AB. The shadow has a width EF = AB = 1 so A_h is (BF − BD) × 1 or $\sin\alpha \cot\beta - \cos\alpha$.

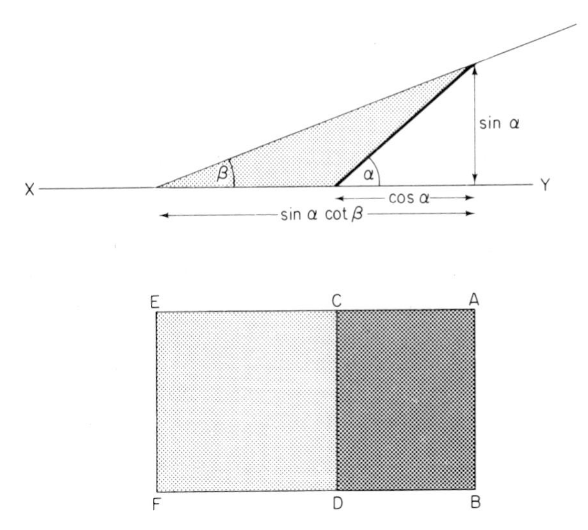

Fig. 4.3 Geometry of rectangle projected on horizontal surface when edge is parallel to solar beam.

If the beam made an angle θ with respect to the direction AC (Fig. 4.4), the shadow would move to the position indicated by CE′F′D where AE′ = AE = sin α cot β. The shadow becomes a parallelogram with an area CG × CD so A_h is (AE′ cos θ − AC) × 1 or {sin α cot β cos θ − cos α}. When β > α, the area is {cos α − sin α cot β cos θ} and, for all values of α and β, the projected area can be written |cos α − sin α cot β cos θ|, the positive value of the function. Because any plane with area A can be subdivided into a large number of small unit squares, the shape factor is

$$\frac{A_h}{A} = |\cos\alpha - \sin\alpha \cot\beta \cos\theta| \qquad 4.3$$

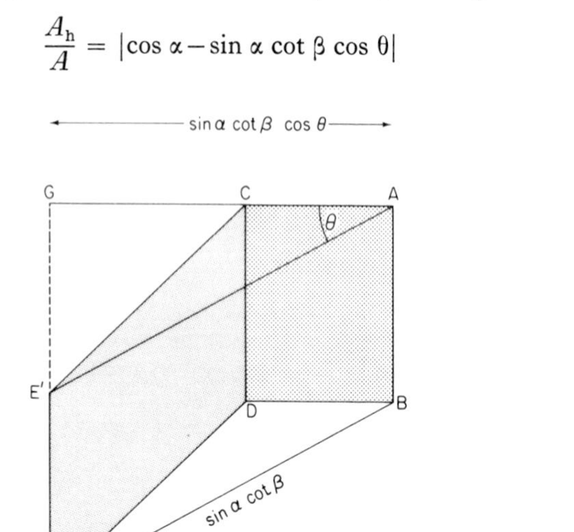

Fig. 4.4 Geometry of rectangle projected on horizontal surface when edge makes angle θ with solar beam.

This function can be used to estimate the direct radiation on hill slopes or on the walls of houses. Then α depends on the geometry of the system, β depends on solar angle, and θ depends both on geometry and on the position of the sun. Relevant calculations and measurements can be found in a number of texts. The example in Fig. 4.5 emphasizes the large difference of direct irradiance on slopes of different aspect[44] which is often responsible for major differences of microclimate and plant response.

Vertical cylinder

Figure 4.6 shows a vertical cylinder of height h and radius r suspended above the ground. The shadow has two components: $h \cot\beta \times 2r$

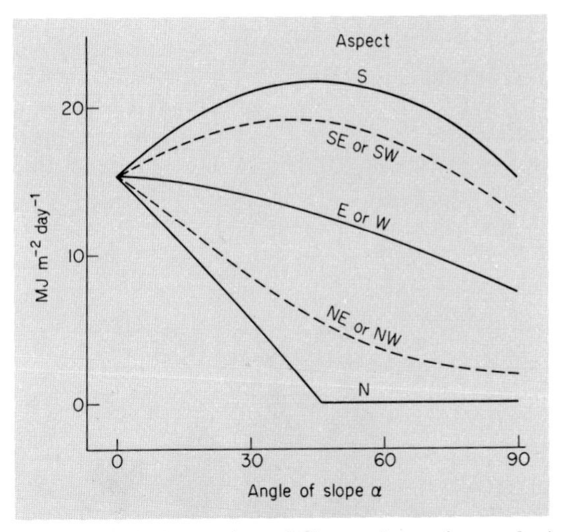

Fig. 4.5 Daily integral of direct solar radiation at the equinoxes for latitude 45°N (from Garnier and Ohmura[44]).

from the curved surface and πr^2 from the top. The total area is $2\pi rh + 2\pi r^2$ so

$$\frac{A_h}{A} = \frac{2rh \cot \beta + \pi r^2}{2\pi rh + 2\pi r^2} = \frac{(2x \cot \beta)/\pi + 1}{2x + 2} \qquad 4.4$$

where $x = h/r$.

Underwood and Ward[147] measured 25 male and 25 female subjects wearing slips or bathing costumes and photographed them from 19

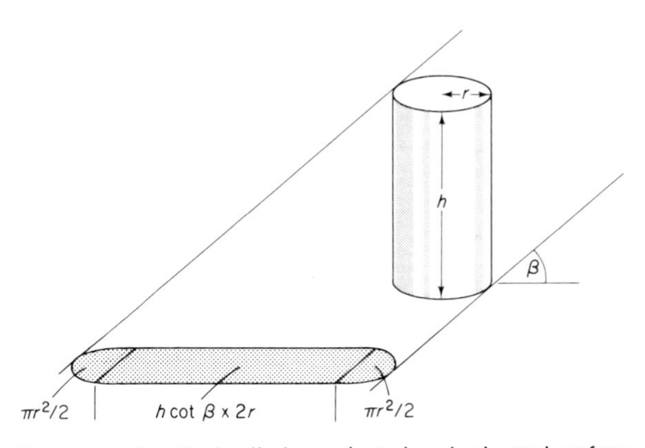

Fig. 4.6 Geometry of vertical cylinder projected on horizontal surface.

different angles. (Silhouettes from 8 angles are shown in Fig. 4.7.) The average projected area of all 50 subjects was found by planimetry of each silhouette with proper correction for parallax. When the areas presented at three azimuth angles (0, 45, 90) were bulked, the mean projected body area was very close to the projected area of a cylinder with

$$h = 1\cdot65 \text{ m}, \quad r = 0\cdot117 \text{ m}, \quad x = 14\cdot1$$

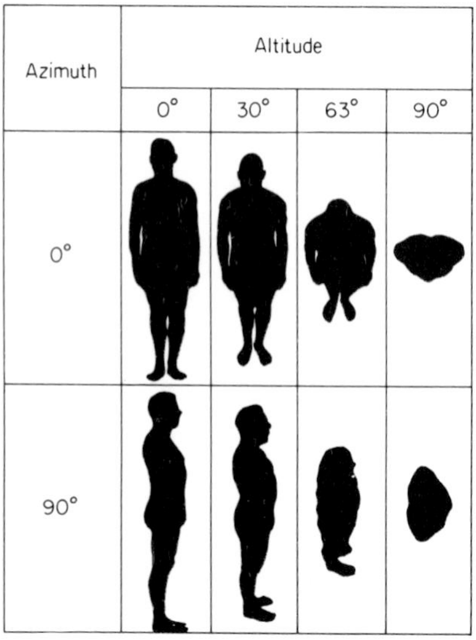

Azimuth	Altitude			
	0°	30°	63°	90°
0°				
90°				

Fig. 4.7 The area of an erect male figure projected in the direction of the solar beam for different values of solar azimuth and altitude. The silhouettes were obtained photographically by Underwood and Ward.

Changes of the projected area with azimuth were taken into account by fitting the measurements to the equation of a cylinder with an elliptical rather than a circular cross section.

Values of A_h/A for vertical cylinders are plotted in Fig. 4.8 as a function of β and x. At a solar elevation of $\beta = 32\cdot5°$, $\cot \beta/\pi = 0\cdot5$ and $A_h/A = 0\cdot5$ independent of x. When $\beta < 60°$ and $x > 10$, A_h/A is almost independent of x, so equation 4.4 with a standard value of $x = 14$ will give a good account of the radiation intercepted by a wide range of human shapes from the ectomorph to the endomorph (e.g. Laurel and Hardy).

Horizontal cylinders

For a horizontal cylinder (h, r), A_h depends on solar azimuth as well as elevation. The azimuth angle θ can be measured from the axis of the

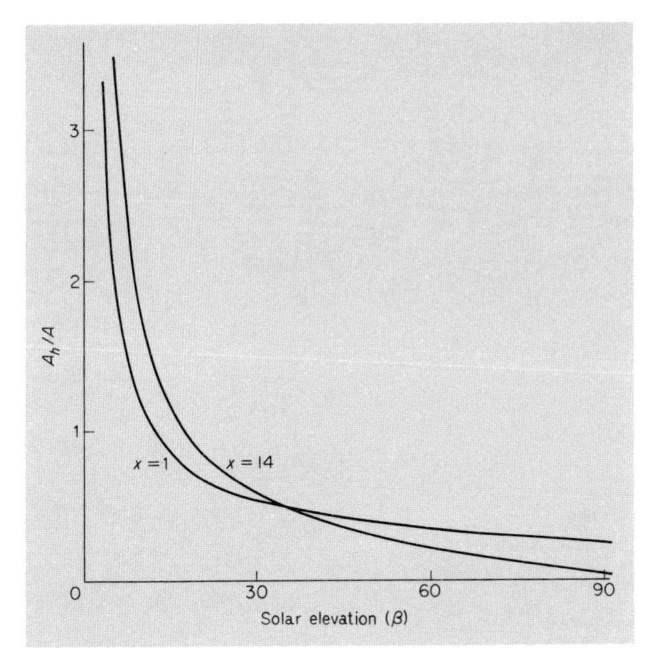

Fig. 4.8 The ratio A_h/A for vertical cylinders with height/radius ratios (x) of 1 and 14.

cylinder pointing in the direction $\theta = 0$. It can be shown that the length h projected in the direction β,θ is $h(1 - \cos^2 \beta \cos^2 \theta)^{1/2}$ and the projected width is $2r$ independent of β and θ. Thus for the *curved surface* only

$$A_h = A_p \operatorname{cosec} \beta = 2rh \operatorname{cosec} \beta(1 - \cos^2 \beta \cos^2 \theta)^{1/2}$$

The illuminated end of the cylinder can be treated as a vertical plane and if α is taken as $\pi/2$ in equation 4.3

$$A_h = \pi r^2 \cot \beta \cos \theta \quad \text{(end only)}$$

The total area of the cylinder including the unlit end is $A = 2\pi rh + 2\pi r^2$ giving for the whole cylinder

$$\frac{A_h}{A} = \frac{2rh \operatorname{cosec} \beta(1 - \cos^2 \beta \cos^2 \theta)^{1/2} + \pi r^2 \cot \beta \cos \theta}{2\pi rh + 2\pi r^2}$$

$$= \frac{2 \operatorname{cosec} \beta\{\pi^{-1}x(1 - \cos^2 \beta \cos^2 \theta)^{1/2} + \cos \beta \cos \theta\}}{2(x+1)} \qquad 4.5$$

For the special case of a cylinder at right angles to the sun's rays, $\theta = \pi/2$, A_h/A reduces to

$$\frac{x \operatorname{cosec} \beta/\pi}{x+1} \quad \text{and} \quad \frac{A_p}{A} = x/\pi(x+1)$$

The projected area of sheep was determined photogrammetrically by Clapperton, Joyce and Blaxter[25] and was compared with the area of

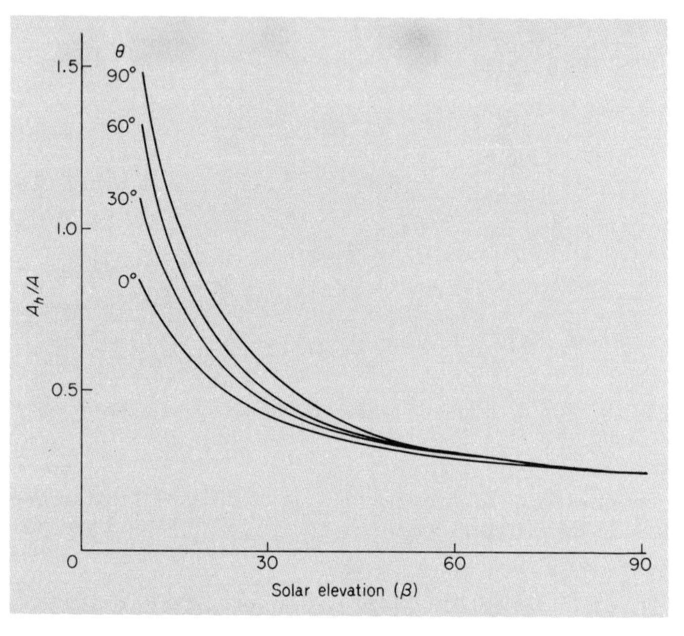

Fig. 4.9 The ratios A_h/A for horizontal cylinders at different values of solar elevation (β) and azimuth (θ) are for a length/radius ratio of $x = 4$.

equivalent horizontal cylinders whose dimensions were determined from photographs at $\theta = 0$ (end view) and $\beta = \pi/2$ (plan view). One set of values quoted for fleeced sheep was

$$h = 0{\cdot}91 \text{ m}, \quad r = 0{\cdot}23 \text{ m}, \quad x = 4{\cdot}1$$

With the sun at right angles to the axis, the cylindrical model underestimated the interception of radiation by about 20% when β was less than 60° but with the sheep facing the sun the model overestimated interception by 10 to 30%. Agreement could probably have been improved by using equation 4.5 instead of calculating h and r from two angles only but, for random orientation, the error in using the dimensions quoted would be very small.

Figure 4.9 shows A_h/A plotted as a function of β for four values of azimuth angle and $x=4$. When β exceeds 40°, the shape factor is almost independent of θ and is therefore nearly proportional to cosec β. This implies that the radiation intercepted by a cylinder with $x=4$ will be almost independent of the solar azimuth provided solar elevation exceeds 40°. For $x<4$ the corresponding angle will be less than 30°.

Cone

The interception of radiation by a cone is an interesting problem, relevant to the distribution of radiant energy on a single tree (Plate 2) or on the leaves of a crop when they are randomly distributed with respect to the points of the compass.

Plate 2 Clipped yew trees (*Taxus baccata*) in the grounds of Hampton Court Palace. The topiarist has produced almost perfect cones casting shadows on the ground which closely resemble the outline of the lower part of Fig. 4.10 (p. 48). The amount of direct solar radiation intercepted by the trees (when they are not shading each other) could therefore be calculated from equation 4.6.

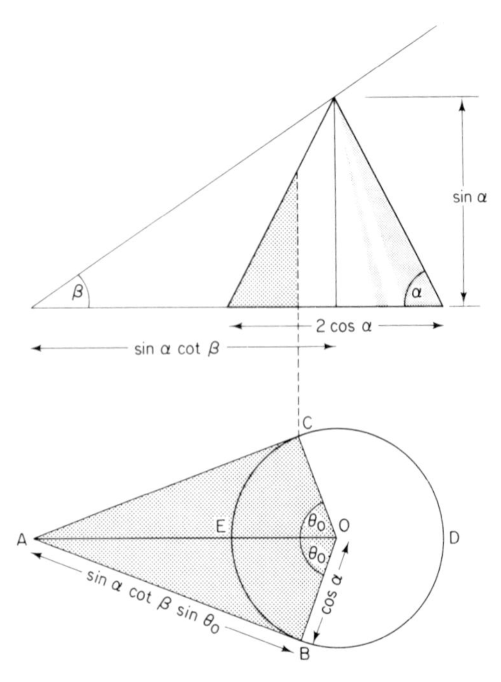

Fig. 4.10 The geometry of a cone projected on a horizontal surface.

The cone in Fig. 4.10 has a slant side of unit length making an angle α with the base, so the perpendicular height is $\sin\alpha$ and the base area is $\pi\cos^2\alpha$. For direct radiation incident at an angle $\beta>\alpha$ (not shown), the walls are fully illuminated and cast a shadow that falls inside the base. The shadow cast on a horizontal surface by the whole cone is simply the shadow area of the base $A_h=\pi\cos^2\alpha$. As the area of the walls is $A=\pi\cos\alpha$, the shape factor for the walls alone is $A_h/A=\cos\alpha$. When $\beta<\alpha$, the shadow has a more complex form. The walls are now partly in shadow: at the base, CDB is illuminated, BEC is not, and the shadow can be delineated by projecting tangents at B and C to meet at A. The sector of the cone in shadow can now be specified by the angle $\text{AOB}=\text{AOC}=\theta_0$. Now $\text{OB}=\cos\alpha$, AO is the horizontal projection of the axis of the cylinder, i.e. $\sin\alpha\cot\beta$, and as ABO is a right angle, $\text{AB}=\sin\alpha\cot\beta\sin\theta_0$. The cosine of θ_0 is $\text{OB}/\text{AO}=\cos\alpha/(\sin\alpha\cot\beta)$. It follows that the area of the shadow is

$$\begin{aligned}\text{ABEC} &= \text{EBDC}+2\text{ABO}-\text{CEBO}\\ &= \text{circle}+2\text{ triangles}-\text{sector of circle}\\ &= \pi\cos^2\alpha+\cos\alpha\sin\alpha\cot\beta\sin\theta_0-\theta_0\cos^2\alpha\\ &= \cos\alpha\{(\pi-\theta_0)\cos\alpha+\sin\alpha\cot\beta\sin\theta_0\}\end{aligned}$$

As the total area of the cone is $\pi \cos\alpha(1+\cos\alpha)$ the shape factor is

$$\frac{A_h}{A} = \frac{(\pi-\theta_0)\cos\alpha+\sin\alpha\cot\beta\sin\theta_0}{\pi(1+\cos\alpha)} \qquad 4.6$$

where $\theta_0=\cos^{-1}(\tan\beta\cot\alpha)$.

DIFFUSE RADIATION

Natural objects are exposed to four discrete streams of diffuse radiation with different directional properties.

(1) Incoming short wave radiation
The spatial distribution of this flux depends on the elevation and azimuth of the sun and on the degree of cloud cover (pp. 27–30).

(2) Incoming long wave radiation
When the sky is cloudless, the intensity of atmospheric radiation decreases by about 20 to 30% from the horizon to the zenith. Under an overcast sky, the flux is nearly uniform in all directions (p. 36).

(3) Reflected solar radiation
The amount of radiation received by reflection from an underlying surface depends on the reflection coefficient of the surface ρ and the angular distribution of the flux is determined by the structure of the surface. Natural vegetation and farm crops are often composed of vertical elements which shade each other and more radiation is reflected from sunlit than from shaded areas.[91]

(4) Long wave radiation emitted by the underlying surface
Like the reflected component of diffuse radiation, the spatial distribution of this flux depends on the disposition of sunlit (relatively warm) and shaded (relatively cool) areas.

The different angular variations of the four diffuse components are difficult to handle analytically but for the purposes of establishing the radiation balance of a leaf or animal they can often be ignored. The following treatment deals with the interception of isotropic radiation, i.e. the intensities of the diffuse fluxes are assumed independent of angle.

Shape factors

Plane surfaces

A *horizontal* flat plate facing upwards receives diffuse fluxes of $\mathbf{S}_d$ and $\mathbf{L}_d$ in the short and long wave regions of the spectrum. The corresponding fluxes on a downward facing surface are $\rho\mathbf{S}_t$ and $\mathbf{L}_u$.

A plate at any angle α to the horizon receives radiation from all four sources on both its surfaces. To find how the irradiance from each of these sources depends on α, the atmosphere and the ground can be replaced by two hemispheres ACO and ACO′ (Fig. 4.11). The plane of the horizon is ABCD and the plane of the flat plate is DEBF making an angle α with the horizontal plane. The irradiance of the plate from the sector of the sky represented by DCBE could be calculated by dividing this sector into a large number of small elements dA and integrating dA times the cosine of the angle between each element and the normal to the surface of the plate. This process can be avoided by the device described on p. 21, i.e. the sector DEBC is projected on to the plane of the plate.

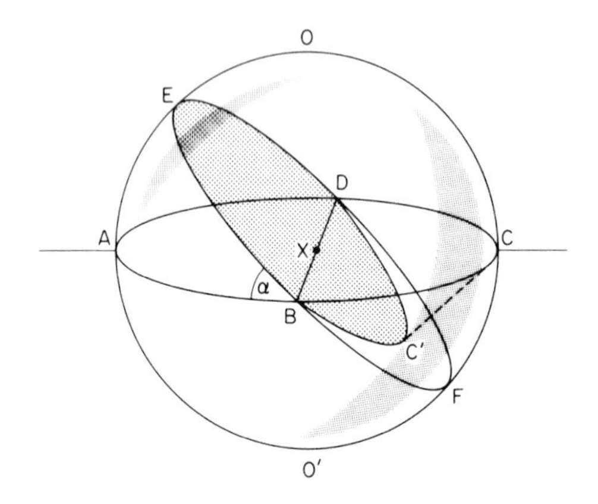

Fig. 4.11 Diagram for calculating the diffuse irradiance at a point X in the centre of a sphere from a sector of the sphere subtending an angle α at the equatorial plane.

The area of this projection is the semi-circle DEBX plus the semi-ellipse DC′BX, i.e. $\pi/2+(\pi \cos \alpha)/2$ if the hemisphere has unit radius. If $\mathbf{N}$ is the (uniform) radiance emitted by elements on the hemisphere, the irradiance from the sector will be $(\pi/2)(1+\cos \alpha)\mathbf{N}$. For a horizontal surface, $\alpha=0$ and the irradiance is $\pi\mathbf{N}$ (p. 21). The irradiance from a surface at angle α is therefore $(1+\cos \alpha)/2=\cos^2(\alpha/2)$ times the irradiance of a horizontal surface. For an inclined plane the factor $(1+\cos \alpha)/2$ for diffuse radiation is equivalent to the factor A_h/A derived for direct radiation.

If a flat leaf or other plane surface is exposed above the ground at an angle α, both its surfaces will receive short and long wave radiation from the sky and from the ground. When the four fluxes of radiation are isotropic, they can be written as follows:

	short wave	long wave
upper surface	$\cos^2(\alpha/2)\,\mathbf{S}_d+\sin^2(\alpha/2)\,\rho\mathbf{S}_t$	$\cos^2(\alpha/2)\,\mathbf{L}_d+\sin^2(\alpha/2)\,\mathbf{L}_u$
lower surface	$\sin^2(\alpha/2)\,\mathbf{S}_d+\cos^2(\alpha/2)\,\rho\mathbf{S}_t$	$\sin^2(\alpha/2)\,\mathbf{L}_d+\cos^2(\alpha/2)\,\mathbf{L}_u$

The sum of all eight components is simply $(\mathbf{S}_d + \rho\mathbf{S}_t + \mathbf{L}_d + \mathbf{L}_u)$. The condition that the upward fluxes of radiation are approximately isotropic requires that the height of the plane above the ground should be large compared with its dimensions so that 'shadow' effect can be neglected.

The total solar radiation received by plane surfaces with different slopes and aspects can be calculated by summing the direct and diffuse components. The curves in Fig. 4.12 were derived by Kondratyev and Manolova[59] who found that it was essential to allow for the spatial distribution of radiation from the blue sky in order to describe the diurnal

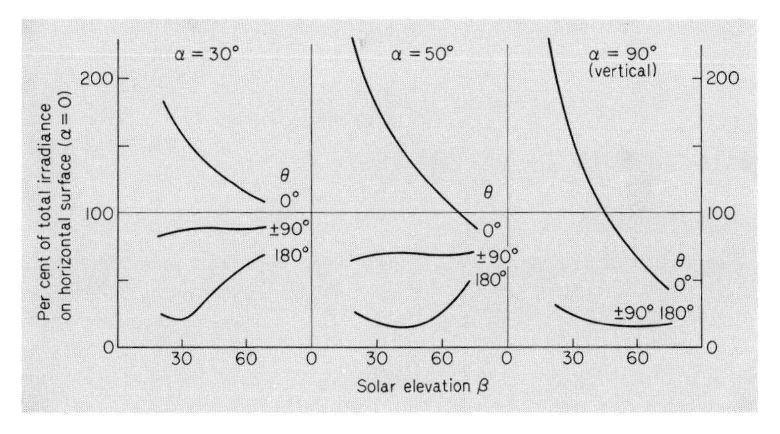

Fig. 4.12 Irradiance of planes (direct and diffuse solar radiation) at latitude 45°N as a function of solar elevation β, elevation of the plane α, and azimuth angle θ between the solar beam and the normal to the plane. From measurements by Kondratyev and Manolova.[59]

change of irradiance on slopes with α exceeding 30°. However, they found that daily totals of radiation on slopes could be calculated accurately by assuming that the diffuse flux was isotropic so that the daily insolation was the sum of the individual hourly values of the direct radiation on the slope, the diffuse flux from the sky $\cos^2(\alpha/2)\mathbf{S}_d$ and the diffuse flux from the surrounding terrain $\sin^2(\alpha/2)\rho\mathbf{S}_t$.

The diffuse irradiance on the walls of a cone with base angle α is equal to the irradiance of the upper surface of a plate at an elevation of α.

Vertical cylinder

For a vertical surface $\cos\alpha = 0$, so the receipt of short wave radiation is $(\mathbf{S}_d + \rho\mathbf{S}_t)/2$ and the receipt of long wave radiation is $(\mathbf{L}_d + \mathbf{L}_u)/2$.

Horizontal cylinder

The components of irradiance for the upper and lower surfaces of a horizontal cylinder can be found by integrating the factors $(1+\cos\alpha)/2$ and $(1-\cos\alpha)/2$. The integration yields factors of $0{\cdot}5+\pi^{-1}\simeq 0{\cdot}82$ and $0{\cdot}5-\pi^{-1}\simeq 0{\cdot}18$. With these approximations, the components are:

	short wave	long wave
upper half surface	$0{\cdot}82\mathbf{S}_d+0{\cdot}18\rho\mathbf{S}_t$	$0{\cdot}82\mathbf{L}_d+0{\cdot}18\mathbf{L}_u$
lower half surface	$0{\cdot}18\mathbf{S}_d+0{\cdot}82\rho\mathbf{S}_t$	$0{\cdot}18\mathbf{L}_d+0{\cdot}82\mathbf{L}_u$

The components for each half surface are the same as they would be for the upper and lower surfaces of a plane with $\alpha\simeq 50°$ and the sum of all the components is simply $(\mathbf{S}_d+\rho\mathbf{S}_t)+(\mathbf{L}_d+\mathbf{L}_u)$ which is the sum for any plane surface.

RADIATION IN LEAF CANOPIES

Direct radiation

The principles of radiation geometry can be applied to estimate the distribution of radiant energy within foliage. The amount of foliage is usually expressed as a **leaf area index** L, the area of leaves above unit area of ground taking only one side of each leaf into account. Suppose a thin horizontal layer in a canopy exposed to direct solar radiation contains a small leaf area index dL. The amount of energy intercepted by dL is the area of shadow cast by the leaves on a horizontal plane times the horizontal irradiance (p. 40). The required shadow area is dL times the shadow cast by unit area of leaf which is (A_h/A). The product $(A_h/A)dL$ is the area of horizontal shadow per unit ground area (a shadow area index) and the intercepted radiation can now be expressed as

$$\begin{aligned} \mathbf{dS}_b &= -(A_h/A)\mathbf{S}_b\, dL \\ &= -\mathcal{K}_s\mathbf{S}_b\, dL \end{aligned} \qquad 4{\cdot}7$$

where

$$\mathcal{K}_s = A_h/A$$

The minus sign is needed if L is measured downwards from the top of the canopy. Integration of this equation gives

$$\mathbf{S}_b(L) = \mathbf{S}_b(0)\, e^{-\mathcal{K}_s L}$$

where $\mathbf{S}_b(L)$ is the direct solar irradiance measured on a horizontal plane below a leaf area index L measured from the top of the canopy. This is a special case of Beer's Law.

The ratio $\mathbf{S}(L)/\mathbf{S}(0)$, which is the relative solar irradiance at L, is also the fractional area of sun flecks on a horizontal plane below L. The area of sunlit foliage[154] between L and $(L+dL)$ is therefore $[\mathbf{S}(L)/\mathbf{S}(0)]\,dL$ and in a stand with a total leaf area of L_t the leaf area index of sunlit foliage is

$$\int_0^{L_t} \mathbf{S}(L)/\mathbf{S}(0)\,dL = \int_0^{L_t} e^{-\mathscr{K}_s L}\,dL = \mathscr{K}_s[1-e^{-\mathscr{K}_s L_t}]$$

which has a limiting value of $1/\mathscr{K}_s$ at large values of L_t. Values of $\mathscr{K}_s$ can be deduced from the area of shadows cast by cylinders, spheres, and cones as presented on pages 40 to 49.

Vertical leaf distribution

If all the leaves in a canopy hung vertically facing at random with respect to azimuth or compass angle, they could be rearranged in a

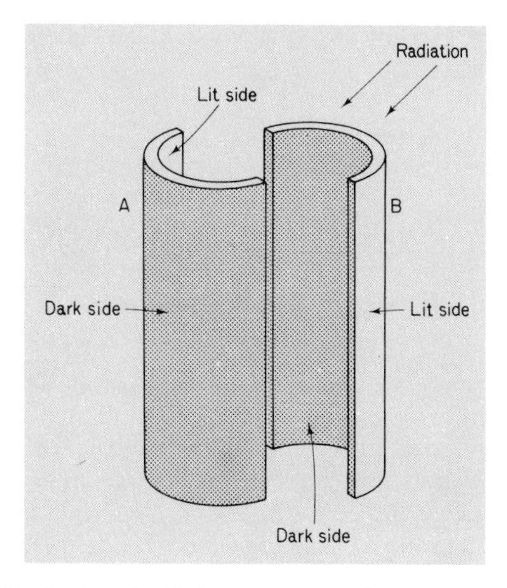

Fig. 4.13 The distribution of radiation over two surfaces of a cylinder representing the irradiance of a large number of vertical leaves.

mosaic pattern on the curved surface of a vertical cylinder. This cylinder could be split along a central plane at right angles to the sun's rays (Fig. 4.13). The convex half of the cylinder represents leaves illuminated on one side (say the upper surface) and the concave half represents the leaves illuminated on the lower surface. The appropriate value of $\mathscr{K}_s = A_h/A$ is therefore twice the value derived for the curved surface of a solid cylinder i.e.

$$\mathscr{K}_s = 2(\cot\beta)/\pi \qquad \text{(cf. p. 43)}$$

Spherical leaf distribution

If the leaves in a canopy were distributed at random with respect to their angles of elevation as well as their azimuth angles, they could be rearranged on the surface of a sphere. Splitting the sphere at the equatorial plane normal to the sun's rays gives two hemispheres representing the two sides from which the leaves could be illuminated. The appropriate value of $\mathscr{K}_s$ is twice the value for a sphere, i.e.

$$\mathscr{K}_s = 0{\cdot}5 \operatorname{cosec} \beta \qquad \text{(cf. p. 40)}$$

Conical leaf distribution

An assembly of leaves all at an elevation of α but distributed at random with respect to their azimuth angles could be rearranged on the curved surface of a cone with a wall angle of α. If the cone is exposed to a solar beam at elevation β, two cases must be considered:

(i) $\beta > \alpha$

All leaves are illuminated from above, i.e. on their upper (adaxial) surfaces. The whole curved surface of the cone is illuminated so

$$\mathscr{K}_s = \frac{A_h}{A} = \frac{\pi \cos^2 \alpha}{\pi \cos \alpha} = \cos \alpha$$

a value which is independent of solar elevation.

(ii) $\beta < \alpha$

Some of the leaves are illuminated from below, i.e. on their lower (abaxial) surfaces. In accordance with the method used for the sphere and the cylinder, the cone can be split into two parts. The relative shadow area of the convex part representing the abaxial surfaces has been calculated already (p. 48): it is $\cos \alpha\{(\pi - \theta_0) \cos \alpha + \sin \alpha \cot \beta \sin \theta_0\}$. The shadow area of the concave part (adaxial surfaces) is the area ACEB in Fig. 4.10, i.e. the sum of two triangles less the sector of the circle or $\cos \alpha\{\sin \alpha \cot \beta \sin \theta_0 - \theta_0 \cos \alpha\}$. The total shadow area is the sum of these expressions and as $A = \pi \cos \alpha$ for the curved surface alone,

$$\mathscr{K}_s = \frac{A_h}{A} = \pi^{-1}\{(\pi - 2\theta_0) \cos \alpha + 2 \sin \alpha \cot \beta \sin \theta_0\}$$

where

$$\theta_0 = \cos^{-1} (\tan \beta \cot \alpha).$$

A similar function, $\mathscr{K}_s \operatorname{cosec} \beta$, which is the relative shadow area on a surface normal to the sun's rays was originally tabulated by Reeve[114] and $\mathscr{K}_s$ has been presented graphically as a function of α and β.[2]

Diffuse radiation

The flux of diffuse radiation within a canopy has two components: radiation from the sky and from clouds which is transmitted through gaps in the foliage, and a diffuse flux generated within the canopy by scattering, i.e. by the reflection and transmission of leaves and other organs. Equations for handling these fluxes have been developed by modifying equation 4.7[62] and also by introducing scattering theory from astrophysics.[118] When the equations are used to estimate rates of photosyntheses, it is important to take account of the spectral distribution of radiation. The fraction of PAR to total radiation is about 0·7 to 0·8 for radiation from the blue sky, and 0·5 for radiation from clouds, but it is only 0·2 for radiation reflected or transmitted by leaves. With increasing depth in a canopy the downward flux of radiation becomes increasingly diffuse and the proportion of infra-red to visible radiation also increases.[131] The change of spectral composition is strongest at about 0·7 μm where the absorptivity of leaves decreases very rapidly with increasing wavelength.

Total irradiance of foliage

To estimate rates of transpiration and photosynthesis for leaves in a canopy, it is essential to calculate the irradiance of individual leaf surfaces as distinct from the irradiance of horizontal surfaces already considered. If $\mathbf{S}_b(L)$ is the direct component of irradiance below an area index of L, the *mean* irradiance of foliage at this depth will be $\mathscr{K}_s\mathbf{S}_b(L)$ from equation 4.7. The mean irradiance can also be derived by considering an irradiance of $\mathbf{S}_b$ (W per m^2 field area) distributed over a sunlit leaf area index of $1/\mathscr{K}_s$ (m^2 leaves per m^2 field area) to give a mean irradiance of $\mathscr{K}_s\mathbf{S}_b$ (W per m^2 leaf area).

In the exceptional case when all leaves are facing in the same direction, the irradiance $\mathscr{K}_s\mathbf{S}_b(L)$ will be uniform but in general some leaves will be exposed to a stronger and some to a weaker flux. In the extreme, leaves parallel to the solar beam ($\alpha=\beta$) receive no direct radiation and leaves at right angles to the beam ($\alpha=\beta+\pi/2$) receive $\mathbf{S}_b$ cosec β. This range of irradiance should be taken into account when estimating maximum leaf temperatures and is relevant to processes which are not proportional to the irradiance, e.g. photosynthesis rates in strong light.[3]

In principle, the diffuse component of irradiance could be determined in the same way as the direct component but a rigorous treatment would need to allow for the change of spatial distribution and spectral composition of the diffuse flux at different depths in the canopy.

Models

The construction of models for estimating the distribution of radiation in plant stands has become a rather fashionable pursuit. In the context of physical ecology, models are groups of equations which describe the environment of an organism and predict its response to environmental factors.[118] Inevitably, many assumptions and simplifications are needed to derive working models of ecological systems and attempts to eliminate assumptions usually increase the complexity of the model and the number of disposable parameters.

Several elementary models of radiation in plant communities can be derived by supposing that a stand with a leaf area index L_t can be divided into nL_t horizontal layers each containing a leaf area index of $1/n$. The salient feature of the foliage in each layer is that the leaves do not overlap, i.e. no leaf can cast a shadow on any other leaf in the same layer. The area of shadow cast on a horizontal surface by the leaves in a sublayer is $\mathscr{K}_s/n$ m^2 shadow per m^2 of ground area. If the transmission coefficient of the leaves is τ, the fraction of radiation transmitted by a sublayer is the sum of the direct radiation penetrating the layer without being intercepted or $(1-\mathscr{K}_s/n)$ and the diffuse radiation generated by interception and forward scattering or $\tau\mathscr{K}_s/n$.

To make progress, it may be assumed that the scattered radiation is intercepted in the same way as direct radiation. Then the radiation transmitted by 2 layers expressed as a fraction of the incident radiation is $(1-\mathscr{K}_s(1-\tau)/n)\times(1\pm\mathscr{K}_s(1-\tau)/n)$ and the fraction transmitted by the whole stand (nL_t layers) is $(1-\mathscr{K}_s(1-\tau)/n)^{nL_t}$. At this point, different treatments diverge. In one model,[88] n is arbitrarily assumed to be unity: there is no overlap in unit leaf area index. Then the transmission of a stand is $\{1-\mathscr{K}_s(1-\tau)\}^{L_t}$, an expression that can also be written in the form $(s+(1-s)\tau)^{L_t}$ where $s=1-\mathscr{K}_s$ and is the fractional area of sun flecks beneath unit leaf layer ($\mathscr{K}_s$ is the fractional area of shadows). Alternatively, if the foliage were distributed at random it would be necessary to make n infinitely large to avoid overlapping of adjacent leaves. The limit of $\{1-\mathscr{K}_s(1-\tau)/n\}^{nL_t}$ as n tends to infinity is $e^{-\mathscr{K}_s(1-\tau)L_t}$, i.e. the case of random foliage corresponds to a conventional Beer's Law type of model. Table 4.1 shows values of s and $\mathscr{K}_s$ determined by measuring the distribution of radiation and leaf area index in plant stands. Values of $\mathscr{K}_s$ for theoretical leaf distributions are given for comparison. To find the mean irradiance of leaves, the horizontal irradiance calculated as a function of L and s (or $\mathscr{K}_s$) is multiplied by $(1-s)$ or $\mathscr{K}_s$.

In principle, it should be possible to determine the value of n from the attenuation of radiation in a stand of known geometry. In practice it is extremely difficult to measure the distribution of leaf area and leaf angle accurately enough to distinguish between the interception of radiation by leaf laminae and by stems, or to measure the direct and diffuse fluxes separately.

From the binomial form of the transmission function, it is possible to estimate how the radiation in a canopy is distributed with respect to levels of irradiance and spectral composition and to allow for corresponding changes of the transmission coefficient of individual leaves. The binomial function is therefore particularly appropriate for calculating photosynthesis rates and it was originally derived for this purpose. The

Table 4.1 Transmission coefficients for model and real canopies[89]

(a) Idealized leaf distributions

	$\mathscr{K}_s$		
	solar elevation β		
	90	60	30
cylindrical	0·00	0·37	1·10
spherical	0·50	0·58	1·00
conical $\alpha = 60$	0·50	0·50	0·58
$\alpha = 30$	0·87	0·87	0·87

(b) Real canopies

	$\mathscr{K}_s$	s
Clover (*Trifolium repens*)	1·10	0·33
Sunflower (*Helianthus annuus*)	0·97	0·38
Beans (*Phaseolus vulgaris*)	0·86	0·42
Kale (*Brassica acephala*)	0·94	0·39
Maize (*Zea mays*)	0·70	0·50
Barley (*Hordeum vulgare*)	0·69	0·50
Beans (*Vicia faba*)	0·63	0·53
Sorghum (*Sorghum vulgare*)	0·49	0·61
Ryegrass (*Lolium perenne*)	0·43	0·65
(*Lolium multiflorum*)		
(*Lolium rigidum*)	0·29	0·75
Gladiolus[96]	0·20	0·85

exponential function may give a better description of the *mean* irradiance at any level as a function of $\mathscr{K}_s$ but it does not predict a distribution of irradiance or of spectral composition. In an extended form, it has been used to estimate the area and irradiance of 'sunlit' and 'shaded' foliage.

Several workers have stressed the prediction that $\mathscr{K}_s$ and s should vary with solar elevation when the leaf elevation α exceeds the solar elevation β[2, 154] (equation 4.6), but the experimental evidence is weak. Indeed most of the measurements that have been made in farm crops show that $\mathscr{K}_s$

and s are nearly constant when β exceeds 30°, i.e. during the central eight hours of the day in the tropics or during the main growing season in temperate latitudes.[89] Constant values of the coefficients can safely be used in estimates of daily photosynthesis or transpiration rates. Models of light transmission are now being developed for canopies in which the arrangement of leaves is not random but the specification of non-randomness leads to much more complex treatments than the methods that have been outlined here.[1,97]

5

Radiation Balance

In precisely the same way that the sun must emit countless particles of light instantaneously, in order that the whole world may constantly be filled with radiance, so all objects must in a moment of time throw off countless idols in countless ways in all directions on every side.

THE EQUATION OF RADIATIVE BALANCE

Radiant energy plays a dominant part in establishing the thermal equilibrium of most plants and animals in their natural environment. The sources of radiation to which organisms are exposed were described in Chapter 3 and geometry for the interception of radiation was considered in Chapter 4. The first part of this chapter reviews the radiative processes which allow organisms to dissipate radiation: reflection and transmission in the short wave spectrum and emission in the long wave spectrum. The second part of the chapter describes the complete radiation balance of natural surfaces, with illustrations for specific cases.

The equation of radiative balance for *unit area* of a surface can be written verbally as

$$\begin{array}{ccccc} \text{BALANCE} & & \text{GAINS} & & \text{LOSSES} \\ \text{Net radiation} & = & \left[\begin{array}{c}\text{incident short}\\ \text{wave radiation}\\ +\\ \text{absorbed long}\\ \text{wave radiation}\end{array}\right] & - & \left[\begin{array}{c}\text{reflected and transmitted}\\ \text{short wave radiation}\\ +\\ \text{emitted long}\\ \text{wave radiation}\end{array}\right] \end{array}$$

With the convention for symbols adopted in the last chapter, the average net radiation per unit area of a body is written $\overline{\mathbf{R}}_n$. The incident short wave radiation consists of direct and diffuse radiation from the sun and the atmosphere, $\bar{\mathbf{S}}_t$, plus sunlight reflected from the environment, $\bar{\mathbf{S}}_e$. The total incident short wave radiation is then $\bar{\mathbf{S}}_t+\bar{\mathbf{S}}_e$ and if the reflection coefficient of the body is ρ_b, the reflected short wave flux is $\rho_b(\bar{\mathbf{S}}_t+\bar{\mathbf{S}}_e)$. Fluxes of long wave radiation to be included in the radiation balance are $\bar{\mathbf{L}}_d$ from the atmosphere, $\bar{\mathbf{L}}_e$ from the environment and $\bar{\mathbf{L}}_b=\sigma \bar{T}_b^{\,4}$, the flux of full radiation at mean surface temperature. A surface with an emissivity of ε will gain $\varepsilon(\bar{\mathbf{L}}_d+\bar{\mathbf{L}}_e)$ *from* its surroundings and emit $\varepsilon\bar{\mathbf{L}}_b$ *to* its surroundings.

The general equation of radiation balance can now be written

$$\mathbf{R}_n = (1-\rho_b)(\bar{\mathbf{S}}_t+\bar{\mathbf{S}}_e)+\varepsilon(\bar{\mathbf{L}}_d+\bar{\mathbf{L}}_e-\bar{\mathbf{L}}_b) \qquad 5.1$$

Before this equation is applied to plants and animals in their natural environments, the main features of their radiative properties will be described.

RADIATIVE PROPERTIES OF NATURAL MATERIALS

All natural materials reflect and transmit solar radiation in the waveband from 0·4 to 3 μm. At the short wavelength, high frequency end of the solar spectrum, the radiative behaviour of materials is determined mainly by the presence of pigments absorbing radiation at wavelengths associated with specific electron transitions. For radiation between 1 and 3 μm, liquid water is an important constituent of many natural materials, because water has strong absorption bands in this region, and even in the visible spectrum where absorption by water is negligible, the reflection and transmission of light by porous materials is often strongly correlated with their water content. In the long wave spectrum beyond 3 μm, most natural surfaces behave like full radiators with absorptivities close to 100% and reflectivities close to zero.

It is important to distinguish between the **reflectivity** of a surface $\rho(\lambda)$, which is the fraction of incident solar radiation reflected at a specific wavelength λ, and the **reflection coefficient** which in this context is the average reflectivity over a specific waveband, weighted by the distribution of radiation in the solar spectrum. If this distribution is described by a function $S(\lambda)$, which is the energy per unit wavelength measured at λ, the total energy in the solar spectrum is $\int_0^\infty S(\lambda)\, d\lambda$. The reflection coefficient of a surface exposed to solar radiation is therefore

$$\bar{\rho} = \frac{\int \rho(\lambda) S(\lambda)\, d\lambda}{\int S(\lambda)\, d\lambda} \qquad 5.2$$

and in practice the integrations may be performed from 0·4 to 3 μm. The transmissivity $\tau(\lambda)$ and the transmission coefficient of a surface can be defined in the same way.

The reflection coefficient of a natural surface is often called the *albedo*, a term borrowed from astronomy and derived from the Greek for 'whiteness'. Because whiteness is associated with the visible spectrum, the wider term 'reflection coefficient' is preferable in this context.

Water

When radiation is incident on clear still water at an angle of incidence ψ less than 45°, the reflection coefficient for solar radiation is almost constant at about 5%. When ψ exceeds 45°, the coefficient increases rapidly with ψ, approaching 100% at grazing incidence (Fig. 5.1). A similar

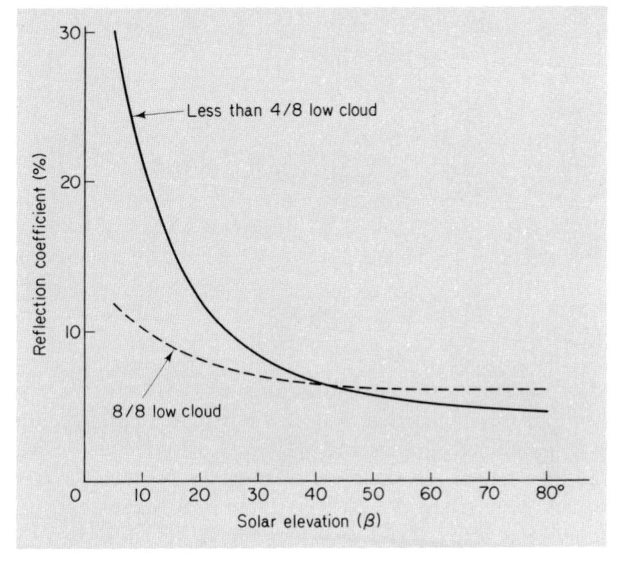

Fig. 5.1 Reflectivity of a plane water surface as a function of solar radiation and cloudiness (from Deacon[30]).

but less pronounced change of reflection with angle is observed from other natural surfaces, e.g. soil and vegetation. The reflectivity of long wave radiation also increases with ψ.[19]

The absorption spectrum of water can be divided into three familiar regions. In the visible spectrum, water is very transparent but red light is absorbed more than blue light. About 11 m of pure water are needed to absorb 10% of incident radiation at 0·65 μm compared with 130 m at 0·45 μm. In the long wave spectrum beyond 3 μm, the absorptivity and the emissivity of water are both about 0·995. Between the visible and long wave regions of the spectrum, water has several absorption bands readily identified in the transmission and reflection spectra of soil, leaves and animal skins. The centres of the main bands are found at 1·45 and 1·95 μm (Fig. 5.2).

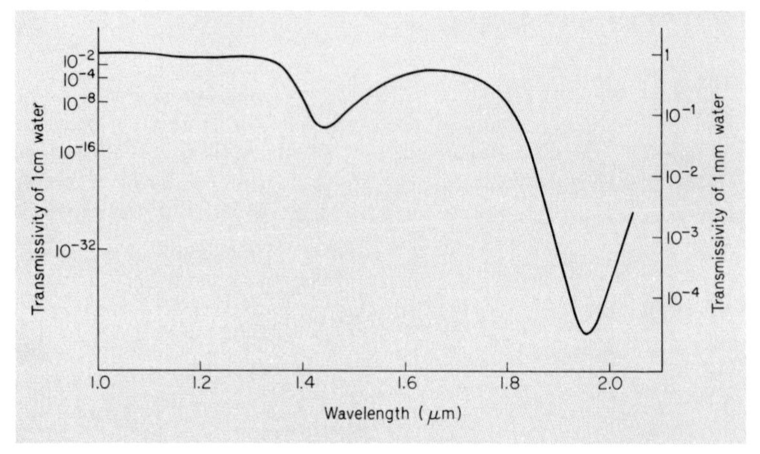

Fig. 5.2 Transmissivity of pure water as a function of wavelength. Note logarithmic scales for 1 cm water (left-hand axis) and 1 mm (right-hand axis).

Soil

The reflectivity of soils depends mainly on their organic matter content, on water content, particle size and angle of incidence. Reflectivity is usually very small at the blue end of the spectrum, increases with wavelength through the visible and near infra-red wavebands and reaches a maximum between 1 and 2 μm. When water is present, its absorption bands are evident at 1·45 and 1·95 μm and in the long wave spectrum many soils have an emissivity between 0·90 and 0·95.

On integration over the whole solar spectrum, reflection coefficients range from about 10% for soils with a high organic matter content to about 30% for desert sand. Even a very small amount of organic matter can depress the reflectivity of a soil. Oxidizing the organic component of a loam, which was 0·8% by weight, increased its reflectivity by a factor of two over the whole visible spectrum.[14]

The reflectivity of clay minerals has been measured as a function of their particle size. Over the spectral range 0·4 to 2 μm, the reflectivity of kaolinite increases rapidly with decreasing particle diameter, e.g. from

56% for 1600 μm particles to 78% for 22 μm particles. Aggregates containing relatively large irregular particles appear to trap radiation by multiple reflection between adjacent faces whereas finely divided powders expose a more uniform surface trapping less radiation. Particle size also governs the transmission of radiation by soils. Baumgartner[6] measured the transmission of artificial light by quartz sand. When the particle diameter was 0·2 to 0·5 mm, a depth of 1 to 2 mm of sand was enough to reduce the radiative flux by 95%, but for particles of 4 to 6 mm, a layer 10 mm deep was needed to give the same extinction. The transmission of radiation by soils has had little attention from ecologists although the effects of light on seed germination and root development are well established.

The reflectivity of a soil sample decreases as it gets wetter, mainly because radiation is trapped by internal reflection at air–water interfaces formed by the menisci in soil pores.[14] The dependence of reflectivity on water content is evident at all wavelengths but is strongest in the absorption band at 1·95 μm. In an example shown in Fig. 5.3, the reflectivity of a loam at 1·9 μm decreased from 60% at 1% water content to 14% at 20% water. The reflection coefficient of a stable soil could therefore be used to monitor the water content of the surface layer.

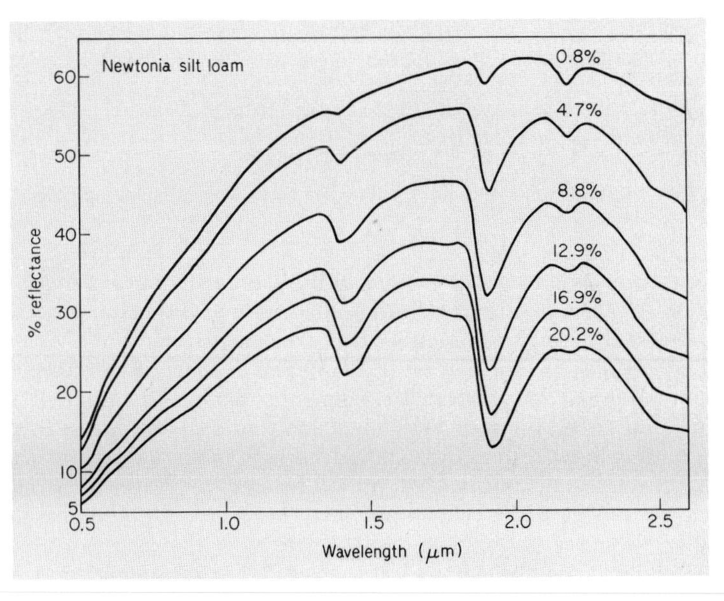

Fig. 5.3 Reflectivity of a loam as a function of wavelength and water content (from Bowers and Hanks[14]).

In the long wave spectrum, most soils have an emissivity between 0·90 and 0·95 and the emissivities of common minerals range from 0·67 for quartz to 0·94 for marble.[19]

Vegetation

The fractions of radiation transmitted and reflected by a leaf depend on the angle of incidence ψ. Measurements by Tageeva and Brandt[133] showed that the reflection coefficient was almost constant for values of ψ between 0 and 50° but as ψ increased from 50° to 90° (grazing incidence), $\rho(\lambda)$ increased sharply as a result of specular reflection. The transmission coefficient was also constant between 0 and 50° but *decreased* between 50

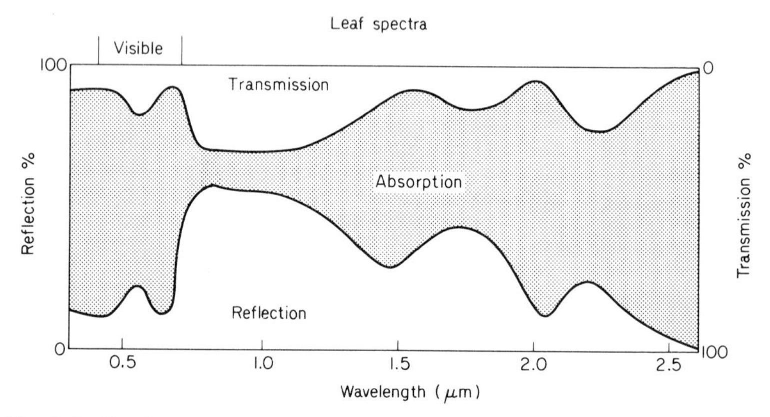

Fig. 5.4 Idealized relation between the reflectivity, transmissivity and absorptivity of a green leaf.

and 90°. Because changes of ρ and τ with angle were complementary, the fraction of radiation absorbed (and available for physiological processes) was almost constant for angles of incidence less than 80°.

In the visible spectrum, most of the radiation penetrating the epidermis is absorbed by chloroplast pigments, notably chlorophyll and carotenoids. Absorption of green light (0·50 to 0·54 μm) is less intense than of blue light (0·40 to 0·47 μm) or red light (0·60 to 0·70 μm) so the light reflected and transmitted by young healthy leaves is strongly green.[46] Over the whole visible spectrum, however, absorption is usually between 80 and 90% provided that the density of chlorophyll exceeds a figure of about 4 mg per m^2 of leaf surface.[43] The remaining 20 to 10% of the incident radiation is scattered in all directions by multiple reflection at cell walls. About the same amount of light is scattered forwards as backwards so the fluxes of visible radiation transmitted and reflected by a leaf are often nearly equal and have a very similar spectral distribution (Fig. 5.4). This implies that the field of visible radiation within a leaf is almost isotropic, i.e. perfectly diffuse, so that individual chloroplasts are exposed to a much smaller flux density than the leaf epidermis. The irradiance needed to saturate the photosynthetic system of many higher plants is therefore much greater than the irradiance at which unicellular algae become light saturated.

Between 0·7 and 1·0 μm, leaf pigments absorb very little radiation, but the absorption by phytochrome at 0·73 μm is of major importance in the control of plant development. Many leaves are almost wholly translucent in this waveband, reflecting and transmitting about 40 to 45% of incident radiation as a result of multiple scattering at cell walls. At wavelengths exceeding 1 μm, spectral behaviour is dominated by the absorption by water and, beyond 3 μm, leaves are almost completely 'black'.

To derive approximate values of ρ and τ for the whole solar spectrum, $\rho(\lambda)$ and $\tau(\lambda)$ can be assumed equal to each other and constant at 0·1 between 0·4 and 0·7 μm and at 0·4 between 0·7 and 3 μm. If each portion of the spectrum contains half of the available radiation, the coefficients are

$$\rho = \tau = (0{\cdot}1 \times 0{\cdot}5) + (0{\cdot}4 \times 0{\cdot}5) = 0{\cdot}25$$

This value is consistent with measurements on a number of species recorded in Table 5.1. Because the visible spectrum makes a relatively small contribution to the reflection and transmission of solar radiation by leaves, comparisons of colour are largely irrelevant to the ranking of reflection coefficients for different species.

The fraction of radiation reflected from the leaf canopy of a field crop or of a natural community is important both in micrometeorology and in ecology. The reflection coefficient of a canopy depends on its geometry and on the angle of the sun as well as on the radiative properties of its components. In general, maximum values of ρ (close to 0·25) are recorded over relatively smooth surfaces such as closely cut lawns. For crops growing to heights of 50 to 100 cm, ρ is usually between 0·18 and 0·25 when ground cover is complete but values as small as 0·10 have been recorded for forests (Fig. 5.5). These differences can be interpreted in terms of the trapping of radiation by multiple reflection between adjacent leaves and stems. For the same reason, the reflection coefficient for most types of vegetation changes with the angle of the sun. Minimum values of ρ are recorded as the sun approaches its zenith and ρ increases as the sun descends to the horizon because there is less opportunity for multiple scattering between the elements of the canopy. The dependence of ρ on zenith angle may explain why reflection coefficients for vegetation

Table 5.1 Radiative properties of plant and animal surfaces

(a) Reflection coefficients ρ (%) for Solar Radiation

1. Leaves

	Ref.	Upper	Lower	Average
Maize (*Zea mays*)	28			29
Tobacco (*Nicotiana tabacum*)	28			29
Cucumber (*Cucumis sativa*)	28			31
Tomato (*Lycopersicon esculentum*)	28			28
Birch (*Betula alba*)	9	30	33	32
Aspen (*Populus tremuloides*)	9	32	36	34
Oak (*Quercus alba*)	9	28	33	30
Elm (*Ulmus rubra*)	9	24	31	28

2. Vegetation—maximum ground cover

(a) Farm crops	Ref.	Latitude of site	Daily mean
Grass	87	52	24
Sugar beet	87	52	26
Barley	87	52	23
Wheat	87	52	26
Beans	87	52	24
Maize	28	43	22
Tobacco	28	43	24
Cucumber	28	43	26
Tomato	28	43	23
Wheat	28	43	22
Pasture	127	32	25
Barley	127	32	26
Pineapple	87	22	15
Maize	98	7	18
Tobacco	98	7	19
Sorghum	98	7	20
Sugar cane	98	7	15
Cotton	98	7	21
Groundnuts	98	7	17
(b) Natural vegetation and forest			
Heather	4	51	14
Bracken	4	51	24
Gorse	4	51	18
Maquis, evergreen scrub	127	32	21
Natural pasture	127	32	25
Derived savanna	98	7	15
Guinea savanna	98	9	19

Table 5.1—continued

(c) Forests and orchards

	Ref.	Latitude of site	Daily mean
Deciduous woodland	4	51	18
Coniferous woodland	4	51	16
Orange orchard	56	32	16
Aleppo pine	127	32	17
Eucalyptus	127	32	19
Tropical rain forest	98	7	13
Swamp forest	98	7	12

3. Animal coats

(a) Mammals

	Ref.	Dorsal	Ventral	Average
Red squirrel (*Tamiasciurus hudsonicus*)	9	27	22	25
Gray squirrel (*Sciurus carolinensis*)	9	22	39	31
Field mouse (*Microtus pennsylvanicus*)	9	11	17	14
Shrew (*Sorex* sp.)	9	19	26	23
Mole (*Scalopus aquaticus*)	9	19	19	19
Gray fox (*Urocyon cinerco argentus*)	9			34
Zulu cattle	12			51
Red Sussex cattle	12			17
Aberdeen Angus cattle	12			11
Sheep weathered fleece	12			26
newly shorn fleece	12			42
Man (*Homo sapiens*)				
Eurasian	12			35
Negroid	12			18

(b) Birds

	Ref.	Wing	Breast	Average
Cardinal (*Richmonenda cardinalis*)	9	23	40	
Bluebird	9	27	34	
Tree swallow	9	24	57	
Magpie	9	19	46	
Canada goose	9	15	35	
Mallard duck	9	24	36	
Mourning dove	9	30	39	
Starling (*Sturnus vulgaris*)	9			34
Glaucous-winged gull (*Larus glaucescens*)	9			52

Table 5.1—continued

(b) Long Wave Emissivities ε (%)

1. Leaves

	Ref.	Average
Maize (*Zea mays*)	53	94·4 ±0·4
Tobacco (*Nicotiana tabacum*)	53	97·2 ±0·6
Snap bean (*Phaseolus vulgaris*)	53	93·8 ±0·8
Cotton (*Gossypium hirsutum* Deltapine)	53	96·4 ±0·7
Sugar cane (*Saccharum officinarum*)	53	99·5 ±0·4
Poplar (*Populus fremontii*)	53	97·7 ±0·4
Geranium (*Pelargonium domesticum*)	53	99·2 ±0·2
Cactus (*Opuntia rufida*)	53	97·7 ±0·2

2. Animals

	Ref.	Dorsal	Ventral	Average
Red squirrel (*Tamiasciurus hudsonicus*)	9	95–98	97–100	
Gray squirrel (*Sciuris carolinensis*)	9	99	99	
Mole (*Scalopus aquaticus*)	9	97	—	
Deer mouse (*Peromyscus sp.*)	9	—	94	
Grey wolf	12			99
Caribou	12			100
Snowshoe hare	12			99
Man (*Homo sapiens*)	12			98

measured in the tropics are usually somewhat smaller than the coefficients for similar surfaces at higher latitudes.

Although it is difficult to distinguish between types of vegetation on the basis of reflection coefficients, differences of reflectivity at particular wavelengths are large enough to allow species to be identified from their spectral signatures. Spectral measurements from aircraft and satellites can now be analysed by computer to produce maps showing the distribution of vegetation over large areas.[26] The distribution of reflected radiation, recorded on infra-red film, can also be used to detect fungal diseases and metabolic disorders which alter the distribution of leaf pigments.[16]

Animals

The coats of furry and hairy animals reflect solar radiation both in the visible and in the infra-red regions of the spectrum. Fig. 5.6 shows that hair behaves rather like soils in the visible spectrum (cf. Fig. 5.3) and that intra- and inter-specific differences of reflectivity are much larger for animals than for vegetation. Reflectivity is therefore an important discriminant in the heat balance of animals but the relation between coat colour and radiative load is complex. The radiation intercepted by the hairs of a coat is scattered forwards, i.e. towards the skin, as well as being reflected away from the animal. The amount of solar radiation reaching the skin will therefore be smaller under a dark coat than under a light coat of the same structure. A dark coat, however, will absorb more

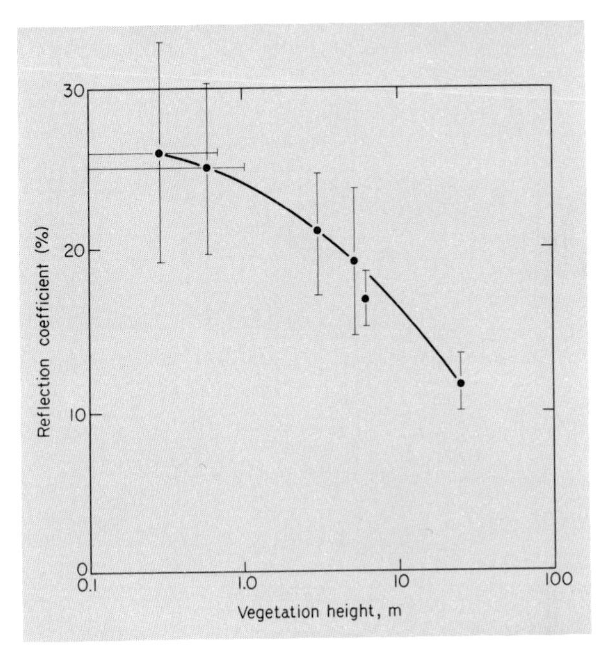

Fig. 5.5 Relation between height of vegetation and reflection coefficient (from Stanhill[127]).

radiation than a light coat, and by maintaining a higher coat temperature will emit more long wave radiation towards the skin. No complete analysis for radiative transfer in coats has yet been published, but a few workers have measured the effective depth to which different types of coat are penetrated by radiation and have shown that the depth of penetration depends on windspeed as well as on coat colour.[52]

The maximum reflectivity of several types of coat is found between 1 and 2 μm. The absorption bands of specific pigments are evident in the

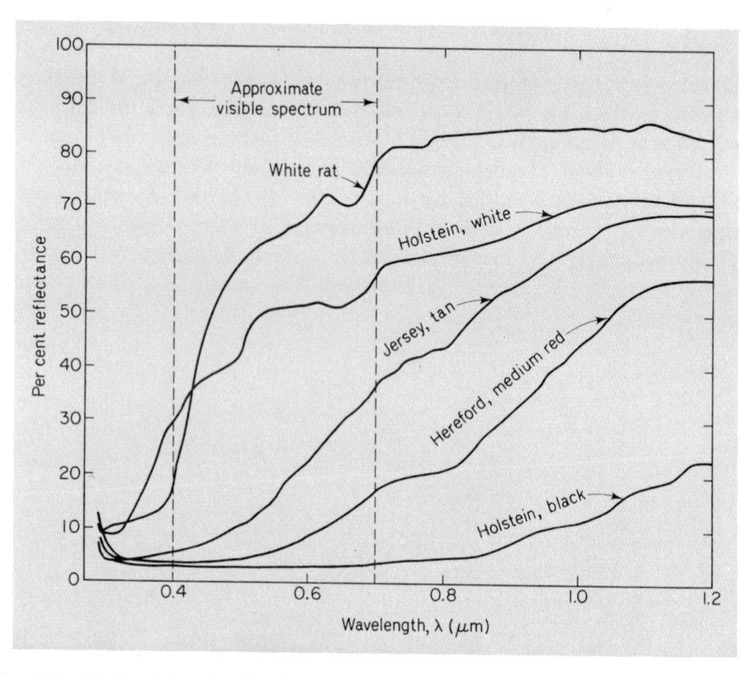

Fig. 5.6 Reflectivity of animal coats (from Mount[95] after Stewart).

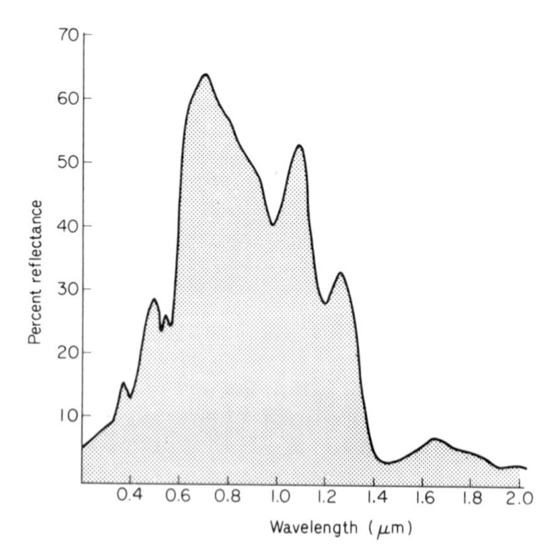

Fig. 5.7 Spectral reflectivity of skin on the author's thumb from a recording by Dr. Warren Porter in his laboratory at the University of Wisconsin on 18 April 1969.

visible spectrum and water is responsible for absorption at 1·45 and 1·95 μm. Beyond 3 μm, the emissivity of coats is usually between 90 and 95%.

In the absence of hair, solar radiation penetrates the skin of animals to a depth that depends on the extent of pigmentation. The depth of penetration in human skin ranges from several millimetres in caucasian subjects with light skins to a few tenths of a millimetre in negroid subjects with a much higher concentration of melanin in the corneum. Corresponding reflection coefficients range from about 20% for dark skins to 40% for light skins. Figure 5.7 shows that the reflectivity of white skin is greatest at about 0·7 μm and decreases to a few per cent at 2 μm.

RADIATION BALANCES

(i) Annual

The radiation balance of natural surfaces will now be examined on three different time scales: one year, one day, and one minute or effectively instantaneously. Figure 5.8 shows the annual change of components in the radiation balance of a short grass surface at Hamburg from February 1954 to January 1955.[41] Each entry in the graph represents the gain or loss of radiation for a period of 24 hours.

The largest term in the balance is $\mathbf{L}_u$, the long wave emission from the grass surface, ranging between winter and summer from about 23 to about 37 MJ m^{-2} day^{-1} (or 270 to 430 W m^{-2}). The minimum values of atmospheric radiation $\mathbf{L}_d$ (230 W m^{-2}) were recorded in spring, presumably in cloudless anticyclonic conditions bringing very cold dry air masses; and maximum values (380 W m^{-2}) were recorded during warm humid weather in the autumn. The net loss of long wave radiation was about 60 W m^{-2} on average (cf. 100 W m^{-2} for cloudless skies on p. 36) and was almost zero on a few very foggy days in autumn and winter.

In the lower half of the diagram, the income of short wave radiation forms a Manhattan skyline with much larger day to day changes and a much larger seasonal amplitude than the income of long wave radiation. The maximum value of $\mathbf{S}_t$ is about 28 MJ m^{-2} day^{-1} (320 W m^{-2}) and $\mathbf{S}_t$ is smaller than $\mathbf{L}_d$ on every day of the year. The reflected radiation is about $0{\cdot}2\mathbf{S}_t$ except on a few days in January and February when snow increased the reflection coefficient to between 0·6 and 0·8.

The net radiation $\mathbf{R}_n$ given by $(1-\rho)\mathbf{S}_t+\mathbf{L}_d-\mathbf{L}_u$ is shown in the top half of the graph. During summer the ratio $\mathbf{R}_n/\mathbf{S}_t$ was almost constant from day to day at about 0·57 but the ratio decreased during the autumn and reached zero in November. From November till the beginning of February $\mathbf{R}_n$ was negative on most days. In summer, net radiation and

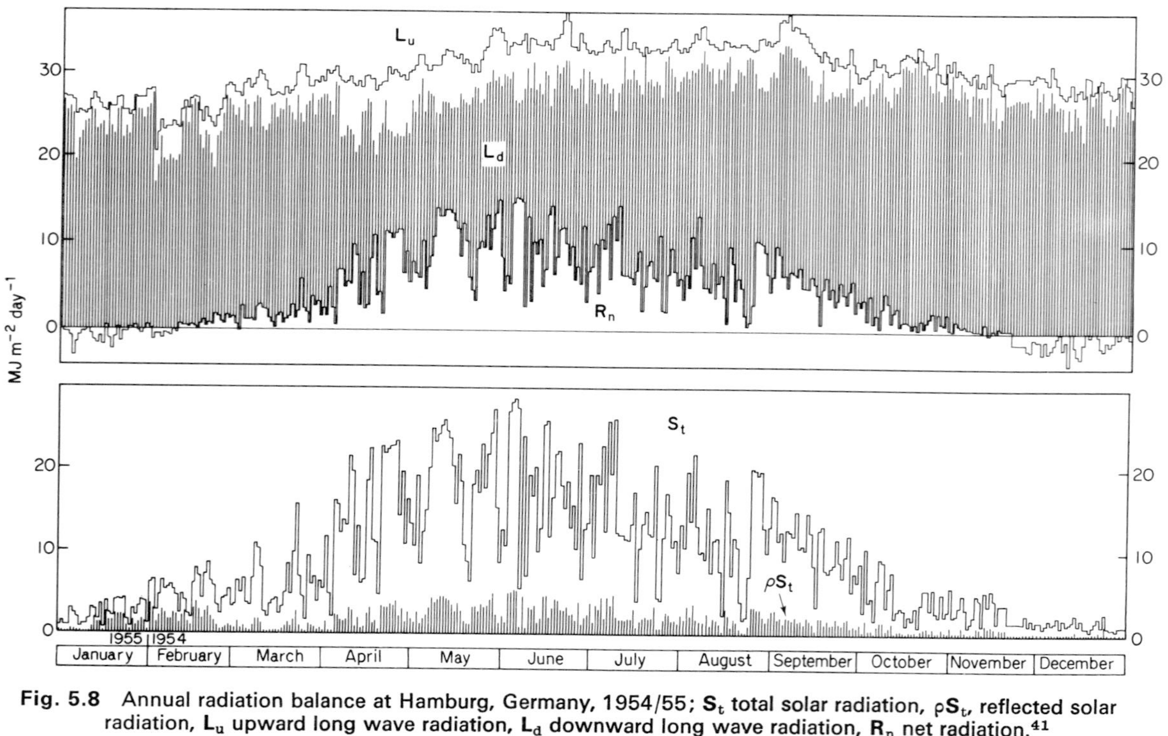

Fig. 5.8 Annual radiation balance at Hamburg, Germany, 1954/55; $\mathbf{S_t}$ total solar radiation, $\rho\mathbf{S_t}$, reflected solar radiation, $\mathbf{L_u}$ upward long wave radiation, $\mathbf{L_d}$ downward long wave radiation, $\mathbf{R_n}$ net radiation.[41]

mean air temperature were positively correlated with sunshine. In winter the correlation was negative: sunny cloudless days were days of minimum net radiation when the mean air temperature was below average.

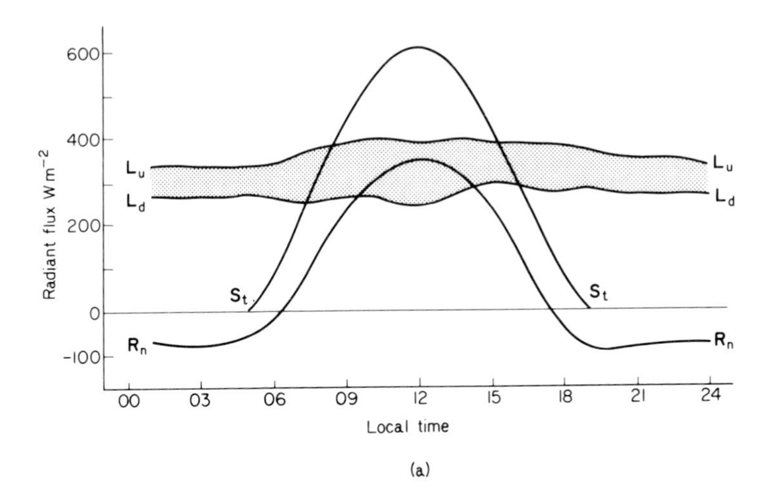

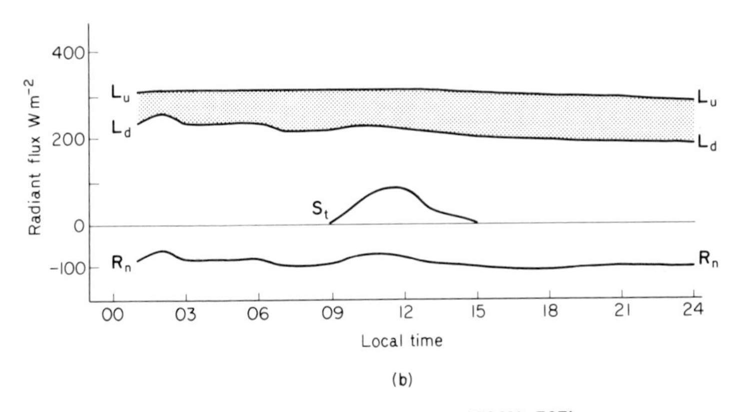

Fig. 5.9 Radiation balance at Bergen, Norway (60°N, 5°E):
(**a**) on 13 April, 1968 (**b**) on 11 January, 1968

The grey area shows the net long wave loss and the line $\mathbf{R}_n$ is net radiation. Note that net radiation was calculated from measured fluxes of incoming short and long wave radiation, assuming that the reflectivity of the surface was 0·20 in April (e.g. vegetation) and 0·70 in January (e.g. snow). The radiative temperature of the surface was assumed equal to the measured air temperature.

(ii) Daily

Components of the radiation balance are recorded continuously at Bergen, Norway, and figures are published by the University of Bergen. Figure 5.9 shows records of (a) total solar radiation $\mathbf{S}_t$, (b) atmospheric radiation $\mathbf{L}_d$ and (c) net loss of long wave radiation from a surface at air temperature, i.e. $\sigma T_a^4 - \mathbf{L}_d$. The net radiation $\mathbf{R}_n$ was calculated for a surface with arbitrary reflection coefficients of 0·2 and 0·7 maintained at air temperature. The shaded parts of the figure represent the loss of short wave radiation by reflection and the *net* loss of long wave radiation; the bold line is $\mathbf{R}_n$. In cloudless weather, the diurnal change in the two long wave components is much smaller than the change of short wave radiation which follows an almost sinusoidal curve. The curve for net radiation is therefore almost parallel to the $\mathbf{S}_t$ curve during the day, decreases to a minimum value in the early evening and then increases very slowly for the rest of the night (because the lower atmosphere is cooled by radiative exchange with the earth's surface). In summer, the period during which $\mathbf{R}_n$ is positive is usually about 2 or 3 hours shorter than the period during which $\mathbf{S}_t$ is positive. Comparison of the curves for clear summer and winter days (Figs. 5.9 a and b) shows that the seasonal change of $\mathbf{S}_t/\mathbf{R}_n$ (Fig. 5.8) is a consequence of (i) the shorter period of daylight in winter and (ii) much smaller maximum values of $\mathbf{S}_t$ in winter, unmatched by an equivalent decrease in the net long wave loss.

In overcast weather, $\mathbf{L}_d$ becomes almost equal to σT_a^4; $\mathbf{R}_n$ is almost zero at night, and during the day $\mathbf{R}_n \simeq (1-\rho)\mathbf{S}_t$.

(iii) Instantaneous

Many of the principles discussed in Chapters 3, 4 and 5 can be illustrated from the instantaneous radiative exchange of contrasting surfaces at different times of day (Fig. 5.10). Four surfaces have been chosen for this exercise:

(i) A short grass lawn.

For a continuous horizontal surface receiving radiation from above and not from below, the net radiation is simply

$$\mathbf{R}_n = (1-\rho_1)\mathbf{S}_t + \mathbf{L}_d - \sigma T_1^4 \qquad 5{\cdot}3a$$

where ρ_1, T_1 are the reflection coefficient and radiative temperature of the lawn.

(ii) A horizontal leaf.

If the leaf is assumed to be exposed above the lawn it receives an additional income of short wave radiation $\mathbf{S}_e = \rho_1 \mathbf{S}_t$ and of long wave radiation $\mathbf{L}_e = \sigma T_1^4$. The net radiation is therefore

$$\mathbf{R}_n = (1-\rho_2)(1+\rho_1)\mathbf{S}_t + \mathbf{L}_d + \mathbf{L}_e - 2\sigma T_2^4 \qquad 5.3b$$

where ρ_2, T_2 are the reflection coefficient and radiative temperature of the leaf. Note that $\mathbf{R}_n$ is the net radiation per unit of *total* leaf area, i.e. twice the plane area or twice the leaf area index. Some workers have calculated the radiation balance of an isolated leaf with respect to the area of a single side but this convention leads to anomalously large values of $\mathbf{R}_n$.

(iii) A sheep.

The sheep is assumed to be standing on the lawn and so receives radiation reflected and emitted by the surface below it. When the area of shadow is ignored the net radiation for the sheep is

$$\mathbf{R}_n = (1-\rho_3)(1+\rho_1)\overline{\mathbf{S}_t} + \overline{\mathbf{L}_d} + \overline{\mathbf{L}_e} - 2\sigma \overline{T_3}^4 \qquad 5.3c$$

where the bars indicate averaging over the exposed surface. The sheep is assumed to be a horizontal cylinder with its axis at right-angles to the sun's rays, reflectivity ρ_3 and mean surface temperature T_3.

(iv) A man.

For a man standing on the lawn, the radiation balance is formally identical to equation 5.3c with ρ_4, T_4 replacing ρ_3, T_3.

Table 5.2 Conditions assumed for radiation balances, Fig. 5.10

	1 High sun clear	2 High sun partly cloudy	3 Low sun clear	4 Overcast day	5 Clear night
Solar elevation β	60	60	10	—	—
Direct solar radiation $\mathbf{S}_b$ (W m^{-2})	800	800	80	—	—
Diffuse solar radiation $\mathbf{S}_d$ (W m^{-2})	100	250	30	250	—
Downward long wave radiation $\mathbf{L}_d$ (W m^{-2})	320	370	310	380	270
Surface temperature (°C)					
air	20	20	18	15	10
lawn	24	24	15	15	6
leaf	24	25	15	15	4
sheep	33	36	15	20	10
man	38	39	15	20	10
Reflectivities					
lawn	0·23	0·23	0·25	0·23	—
leaf	0·25	0·25	0·35	0·25	—
sheep	0·40	0·40	0·40	0·40	—
man	0·15	0·15	0·15	0·15	—

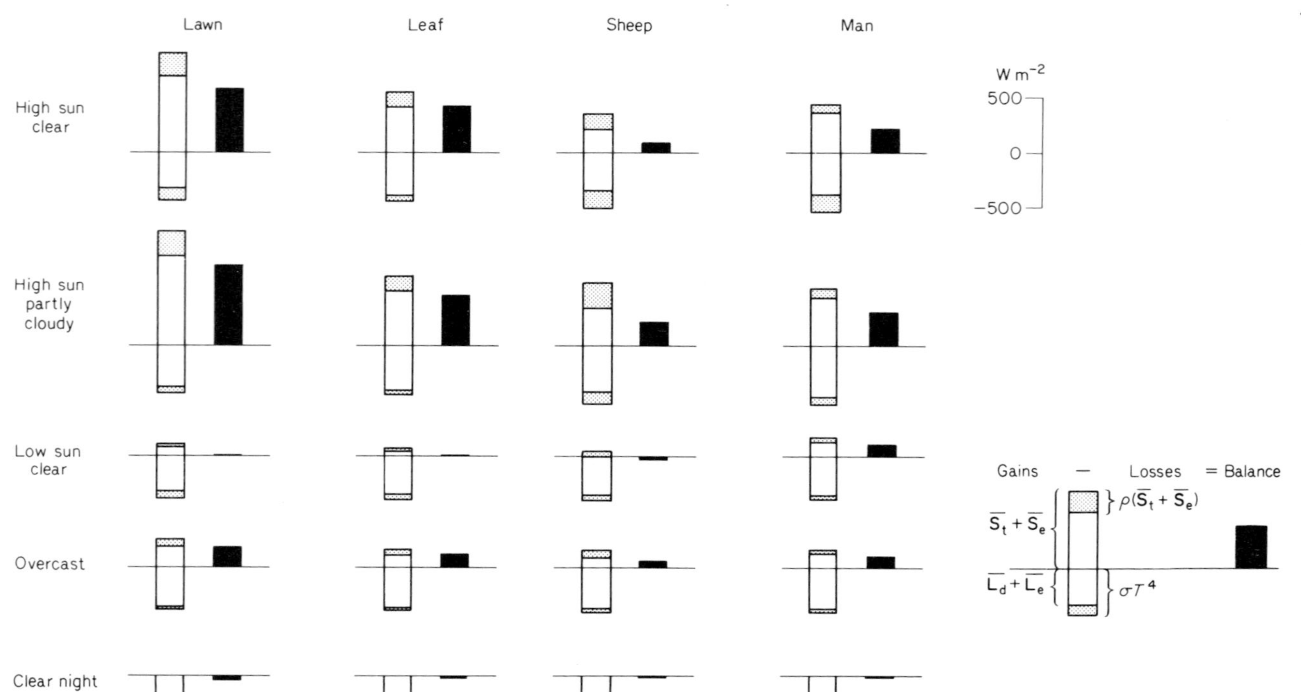

Fig. 5.10 Radiation balance for different surfaces in a range of weather conditions as specified in Table 5.2.

Values assumed for the radiation fluxes and for ρ and T are given in Table 5.2. Long wave emissivities are assumed to be unity.

It is instructive to compare the net radiation received by different surfaces at the same time, noting the effects of geometry and reflection coefficients; or to compare the same surface at different times to see the influence of solar elevation and cloudiness. Salient features of these comparisons are as follows.

(a) The lawn absorbs more net radiation than any of the other surfaces including the isolated leaf. The leaf receives less short wave radiation than the lawn ($(1+\rho_1)\mathbf{S}_t/2$ compared with $\mathbf{S}_t$) and absorbs more long wave radiation ($(\mathbf{L}_d+\sigma T_1{}^4)/2$ compared with $\mathbf{L}_d$).

(b) The sheep absorbs less net radiation than the other surfaces. This is partly a consequence of the relatively large reflection coefficient (0·4) and partly a consequence of geometry.

(c) The geometry of the man ensures that $\mathbf{R}_n$ is large in relation to other surfaces when the sun is low.

(d) For all surfaces, the net radiation is greatest when the sun is shining between clouds and is larger under an overcast sky than it is when the sun is near the horizon.

(e) At night, the leaf, sheep and man receive long wave radiation from the lawn as well as from the sky so their net loss of long wave radiation is less than the net loss from the lawn.

6

Momentum Transfer

The wild wind awakened whips the waves of the sea, capsizes huge ships, and sends the clouds scudding; sometimes it swoops and sweeps across the plains in tearing tornado, strewing them with great trees, and hammers the heights of mountains with forest-splitting blasts.

When plants or animals are exposed to radiation, the energy which they absorb can be used in three ways: for heating, for the evaporation of water, and for photochemical reactions. Heating of the organism itself or of its environment implies a transfer of heat by conduction or by convection; evaporation involves a transfer of water vapour molecules in the system and photosynthesis involves a similar transfer of carbon dioxide molecules. At the surface of an organism, heat and mass transfer are sustained by molecular diffusion through a thin skin of air known as a **boundary layer** in contact with the surface. The behaviour of this layer depends on the viscous properties of air and on the transfer of momentum associated with viscous forces. A short discussion of momentum transfer is therefore needed as background to the following three chapters which consider different aspects of exchange between organisms and their environment.

BOUNDARY LAYERS

Figure 6.1 shows the development of a boundary layer over a smooth flat surface immersed in a moving fluid (i.e. a gas or liquid). When the streamlines of flow are almost parallel to the surface the layer is said to be **laminar** and the flow of momentum across it is effected by the momentum exchange between individual molecules discussed on pp. 4–5. The thickness of a laminar boundary layer cannot increase indefinitely because the flow becomes unstable and breaks down to a chaotic pattern of swirling motions called a **turbulent** boundary layer. A second laminar layer of restricted depth may start to form within the turbulent layer. Both in laminar and in turbulent layers, velocity increases with distance from the surface, but not in a simple linear way.

Because there is no sharp discontinuity of velocity between a boundary layer and the free airstream, definitions of boundary layer depth are necessarily rather arbitrary. Depth can be defined by the streamline along which the velocity reaches 99% of its value in the free stream but, in problems of exchange, it is more convenient to work with an average boundary layer depth.

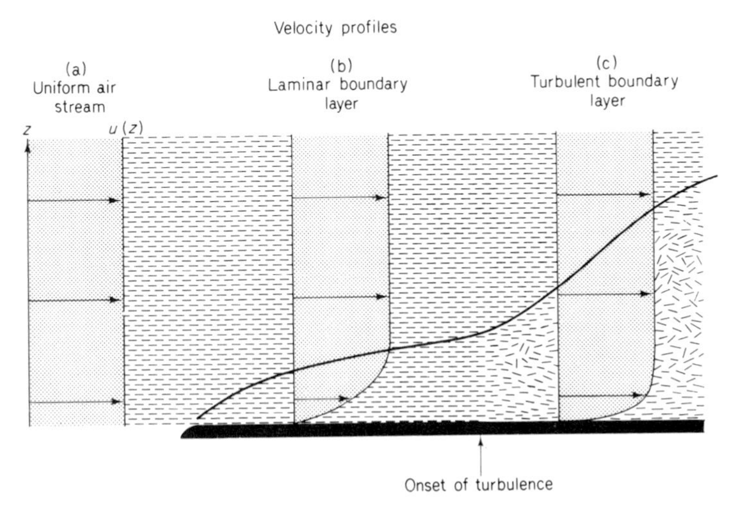

Fig. 6.1 Development of laminar and turbulent boundary layers over a smooth flat plate. (The vertical scale is greatly exaggerated.)

In Fig. 6.2, the flow of air reaching a flat plate across a vertical cross-section of depth h is proportional to the velocity V times h. At a distance l from the edge of the plate, the velocity profile is represented by the line ABC and the flow through the cross section at C will be less than Vh because the velocity in the boundary layer is less than V. The same reduction in flow would be produced by a layer of completely still air with thickness δ (shaded) above which the air moves with a uniform velocity V. The velocity profile in this equivalent system is represented by ADFC. The depth δ is known as the 'displacement boundary layer' and can be regarded as an average depth of the boundary layer between the leading edge of the plate and the cross section at C.

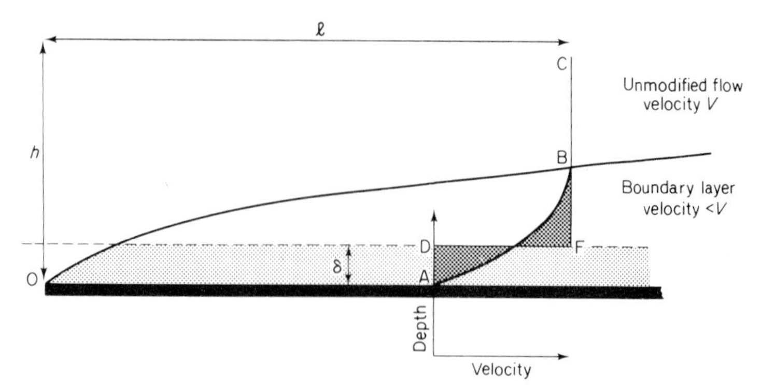

Fig. 6.2 Boundary layer OB, displacement boundary layer (grey) and wind profile CBA over a smooth flat plate exposed to an airstream with uniform velocity V.

By applying the principle of momentum conservation to flow within a laminar boundary layer, it is possible to show that the displacement thickness increases with the square root of the distance l. The relation

$$\delta = 1{\cdot}72(l\nu/V)^{1/2} \tag{6.1}$$

shows how the depth also depends on the kinematic viscosity of air ν and on the velocity V. Division of both sides of equation 6.1 by l gives the ratio of depth to length

$$\delta/l = 1{\cdot}72(Vl/\nu)^{-1/2} \tag{6.2}$$

Both sides of this equation are non-dimensional and the quantity Vl/ν is called a **Reynolds number**, ubiquitous in fluid dynamics and named after a nineteenth century physicist. A Reynolds number, often written Re, expresses the ratio of inertial forces in a fluid (producing changes in velocity) to viscous forces (tending to oppose changes of velocity). When the ratio is small, viscous forces predominate so that the flow tends to remain laminar, but when the ratio increases beyond a critical value, inertial forces dominate the system and the flow becomes turbulent. For a smooth flat plate in a wind tunnel exposed to air with little inherent turbulence, the critical value of Re is about 3×10^5 so, for a velocity of 1 m s^{-1}, the critical depth is 1·4 cm achieved at a distance of $l = 4{\cdot}5$ m.

Skin friction

The force that air exerts tangentially as it flows over a surface is a direct consequence of momentum transfer through the boundary layer and is known as **skin friction**. To establish analogies with heat and mass transfer later, the transfer of momentum can be treated as a diffusion process (see p. 12). If t is a diffusion pathlength for the transfer of momentum from air moving with velocity V to a surface where the velocity is zero, the appropriate velocity gradient is V/t. Then the frictional force is

$$\tau = \nu\rho V/t \tag{6.3}$$

From theoretical analysis of the flow in a laminar boundary layer, the drag per unit area on a smooth flat surface is proportional to $V^{3/2}$ and is given by

$$\tau = 0{\cdot}664\nu\rho V(V/l\nu)^{1/2}$$

Substituting for l from equation 6.2 gives

$$\tau = 1{\cdot}14\nu\rho V/\delta \tag{6.4}$$

Comparison of equations 6.3 and 6.4 shows that $t = 0{\cdot}88\delta$, i.e. the rate of momentum transfer across a real boundary layer with a thickness proportional to $l^{1/2}$ is nearly the same as the rate of transfer across the fictitious layer of still air shown with a thickness of δ in Fig. 6.2. As the diffusion resistance r_M is t/ν (p. 14), equation 6.1 and the relation $t = 0{\cdot}88\delta$ can be combined to give

$$r_M = 1{\cdot}5(l/V\nu)^{1/2} \tag{6.5}$$

For example, if $V = 1$ m s^{-1} and $l = 5$ cm, $r_M = 0{\cdot}9$ s cm^{-1}, establishing an order of magnitude for many micrometeorological problems.

Form drag

In addition to the force exerted by skin friction, a consequence of momentum transfer to a surface across the streamlines of flow, bodies immersed in moving fluids experience a force in the direction of the flow as a result of the deceleration of fluid. This force is known as form drag because it depends on the shape and orientation of the body. Maximum form drag is experienced by surfaces at right angles to the fluid flow and the force can be estimated by assuming that there is a point on the surface where the fluid is instantaneously brought to rest after being uniformly decelerated from a velocity V. If the initial momentum per unit volume of fluid is ρV and the mean velocity during deceleration is $V/2$, the rate at which momentum is lost from the fluid is $\rho V \times V/2 = \frac{1}{2}\rho V^2$. This is the maximum rate at which momentum can be transferred to unit area on the upstream surface of a bluff body and it is therefore the maximum pressure excess that a fluid can exert in contributing to the total form drag over the body. In practice, fluid tends to slip round the sides of a bluff body so that a force smaller than $\frac{1}{2}\rho V^2$ is exerted on the upstream face.

However, in the wake which forms downstream of a bluff body, the fluid pressure is less than in the free stream and the associated suction often makes an important contribution to the total form drag on the body. The total drag force on unit area is conveniently expressed as $c_f . \frac{1}{2}\rho V^2$, where c_f is a **drag coefficient**.

In most problems, it is appropriate to combine skin friction and form drag to give a total force τ, usually the force on a unit of area projected in the direction of the flow, i.e. $2rl$ for a cylinder of radius r and length l in cross-flow and πr^2 for a sphere. The ratio $\tau/\frac{1}{2}\rho V^2$ then defines the total drag coefficient c_d which for spheres and for cylinders at right angles to the flow lies between 0·4 and 1·2 in the range of Reynolds numbers between 10^2 and 10^5. The drag coefficient of a thin flat plate is discussed in the next section.

As background for later discussion of mass and heat transfer, it should be noted that the diffusion of momentum in skin friction is analogous to the diffusion of gas molecules and of heat provided the surface is parallel to the airstream. For such a surface, close relationships may be expected between r_M, r_H and r_V. For a surface at right angles to the airstream, however, there is no *frictional* force in the direction of flow. Friction will operate in all directions at right angles to the flow but the net sum of all these (vector) forces must be zero. Similarly, the net flux of heat or mass will be zero in the plane of the surface but must be finite in the direction of the flow. In this case, r_V and r_H may be similar to each other but will be unrelated to the value of r_M.

Drag on natural surfaces

The atmosphere in motion exerts forces on all natural surfaces—individual leaves, plants, trees, crops, animals, bare soil and open water. Conversely, every object or surface exposed to the force of the wind imposes an equal and opposite force on the atmosphere proportional to the rate of momentum transfer between the air and the surface. Momentum transfer is always associated with wind 'shear': the windspeed is zero at the surface of the object and increases with distance from the surface through a boundary layer of retarded air.

Isolated objects like single plants or trees tend to have very irregular boundary layers and disturb the motion of the atmosphere by setting up a train of eddies in their wake rather like the eddies formed downstream from the piers of a bridge. Surfaces like bare soil and uniform vegetation also generate eddies in the air moving over them because the drag which they exert on the air is incompatible with laminar flow. Over extensive level surfaces of this kind, the eddies constitute a turbulent boundary layer in which the windspeed increases with the logarithm of height above the surface. From the shape of this wind profile, it is possible to determine the aerodynamic 'roughness' of the underlying surface and to establish a relation between the speed and force of the wind.

Drag on leaves

To avoid the fluctuations of windspeed that are characteristic of the atmosphere the drag on natural objects can be measured in a wind tunnel where the flow is steady and controlled. Thom[137] studied the force on a

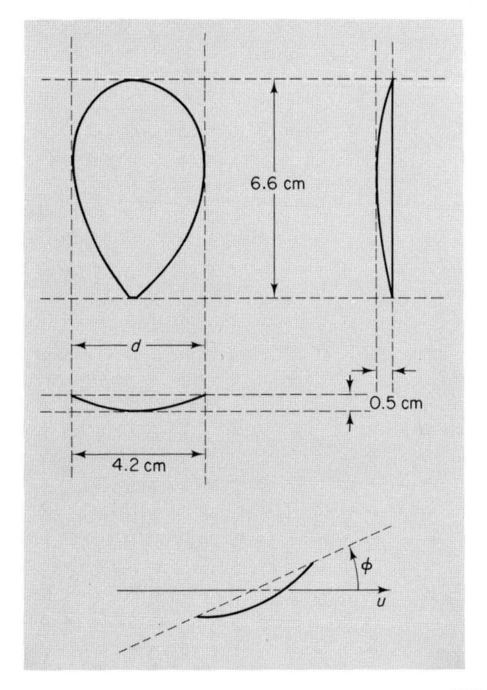

Fig. 6.3 Shape and dimensions of model leaf used by Thom.[137]

replica of a 'leaf' made of thin aluminium sheet. Figure 6.3 shows the dimensions of the replica and Fig. 6.4 shows how the drag coefficient c_d changed with windspeed and direction. Note that the quantity c_d shown in Fig. 6.4 was calculated by dividing the force per unit *plan area* of the leaf (τ/A) by ρV^2 which is numerically the same as the force per unit of total surface area ($\tau/2A$) divided by $\frac{1}{2}\rho V^2$. This departure from an aerodynamic convention (which uses a projected area) allows the resistance to momentum transfer for the whole leaf to be written as $r_M = 1/Vc_d$, equivalent to the combination in parallel of the two resistances of $2/Vc_d$ for each surface separately.

When the leaf was oriented in the direction of the airstream ($\phi = 0$) the drag was minimal and c_d was close to twice the theoretical value derived from equation 6.5 which gives the resistance to momentum transfer for one side of a plate only. When the concave or convex surfaces were facing the airstream ($\phi = +90$ or $\phi = -90$ respectively) form drag was much larger than skin friction.

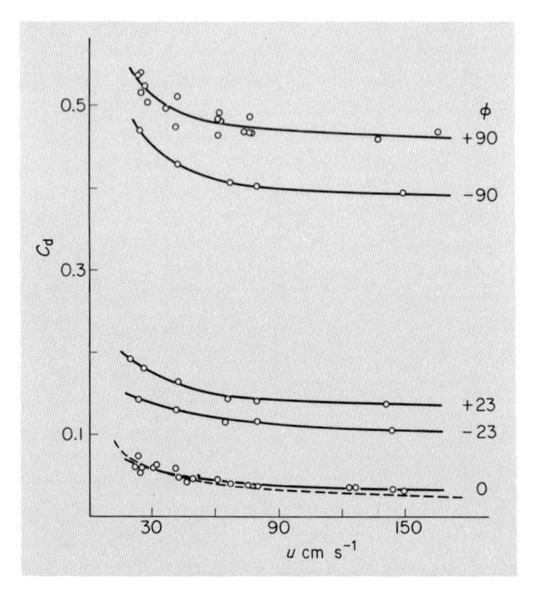

Fig. 6.4 Drag coefficient of a model leaf (full lines) as a function of windspeed u and angle between leaf and airstream 0 (see Fig. 6.2). The broken curve is a theoretical relationship for a thin flat leaf at $\phi = 0$.

The general form of the curves in Fig. 6.4 is consistent with a combination of form drag proportional to V^2 and skin friction proportional to $V^{3/2}$. The total drag coefficient c_d can be expressed as the sum of a form drag component c_f and a frictional component $nV^{-1/2}$ where n is a constant, i.e.

$$c_d = c_f + nV^{-1/2} \tag{6.6}$$

The relevance of wind tunnel measurements of leaf drag to momentum transfer in a crop is a matter for debate. In the first place, turbulence is usually suppressed in wind tunnels to achieve laminar flow whereas the movement of air in canopies is nearly always turbulent. Turbulent eddies of an appropriate size can increase rates of momentum transfer by disturbing flow in the laminar boundary layer, and the disturbance is likely to be accentuated if the turbulence is strong enough to make the leaves flutter. Measurements by Roshko indicate that the characteristic diameter of eddies shed by cylinders of diameter d is about $5d$ when Re exceeds 200. The eddies shed by leaves and stems decay to form smaller and smaller eddies drifting downwind and penetrating the boundary layer of other leaves and stems.

The drag on real leaves may also be increased by the roughness of the cuticle and by the presence of hairs. Sunderland[129] found that the drag on the aluminium replica of a wheat leaf increased when a real leaf was attached to the metal surface. The increase was about 20% at 1·5 m s^{-1} growing to 50% at 0·5 m s^{-1} because of the increasing importance of skin friction at low windspeeds.

Lodging of grass and cereal crops occurs when plant stems are unable to withstand the combined forces of wind and of gravity on the foliage and on the ear. Lodging is most likely to occur when the weight of the upper parts of the plants is increased by the interception of rain or when the lower parts of the stems are weakened by disease or by a heavy application of nitrogenous fertilizer. Very little information is available in the literature about the forces needed to bend plant stems beyond this elastic limit but Tani[135] provided figures for rice which indicate the scale of forces which are likely to lodge other cereal crops of similar height, i.e. about 1 m.

Tani found that when mature rice plants were exposed to a uniform wind in the laboratory, the stems broke when the forces on the plants produced a moment of about 0·2 N m on the base of the stems. Plants growing to a height of about 0·84 m in the field were expected to lodge when the windspeed exceeded 20 m s^{-1}. At this speed, the moment on the base of the stem had two components: a moment of 0·034 N m induced by the force of the wind (mainly form drag), and a moment of 0·023 N m induced by the force of gravity acting because the top of the stem was displaced by about 40 cm. The total moment needed to break the stem under field conditions was therefore 0·056 N m, about a quarter of the laboratory value. The discrepancy between the figures can be explained by (a) the much larger forces exerted in the field during strong gusts when the instantaneous windspeed can be two or three times larger than the mean, or (b) a resonance set up between the natural period of oscillation of the plants (about 1 s) and the dominant period of turbulent eddies at the top of the canopy, or (c) the effect of disease on field-grown plants.

Drag on trees

Trees are uprooted in gales when the total bending moment exerted by the trunk on the root system exceeds the maximum restoring movement that the surrounding soil can exert on the roots. Fraser[42] exposed fully grown conifers in a wind tunnel with a diameter of 7 m and found that the relation between drag force and windspeed was strongly affected by a decrease in the effective cross section of the crown as the wind got stronger, a result of streamlining by individual leaves as well as by whole branches (Plate 3). Whereas, for rigid objects, the drag at high windspeeds is almost proportional to V^2, the drag on specimens of four species was found to increase almost linearly with velocity, an important result of streamlining.

The drag coefficient c_d was defined as the total drag force divided by $\frac{1}{2}\rho V^2 A_0$ where A_0 was the area of the crown in *still* air projected in the direction of the airstream. With this convention, c_d at 15 m s^{-1} ranged from 0·57 for spruce (*Picea abies*) to 0·25 for Western hemlock (*Tsuga heterophylla*).

As a test of the validity of the wind tunnel measurements in natural conditions, forest trees were uprooted with a hand winch using a dynamometer to measure the force exerted. Specimens of Douglas fir (*Pseudotsuga taxifolia*) growing to sixty feet (18 m) were uprooted by moments of 5 to 8×10^4 N m with respect to the base of the trees. In the wind tunnel, a windspeed of 25 m s^{-1} was needed to produce this moment. Later, similar trees were uprooted by a gale in which the mean gust speed was close to 25 m s^{-1}.

Having established a valid relation between drag and wind velocity for real trees, Fraser extended his study to the behaviour of small scale forest models and the measurement of bending moments on model trees in different patterns of planting.

WIND AND TURBULENCE OVER UNIFORM SURFACES

The formation of laminar and turbulent boundary layers over a smooth flat plate was described on p. 80. In principle, a laminar boundary layer could form over very smooth natural surfaces like calm water or mud flats, but forests, farm crops and even closely cut lawns are aerodynamically 'rough' even at low windspeeds, in the sense that turbulence is continuously generated by the passage of air over the individual elements of vegetation. In this context, 'rough' implies that the forces retarding the flow of air at the surface are much larger than the viscous forces which tend to maintain laminar flow.

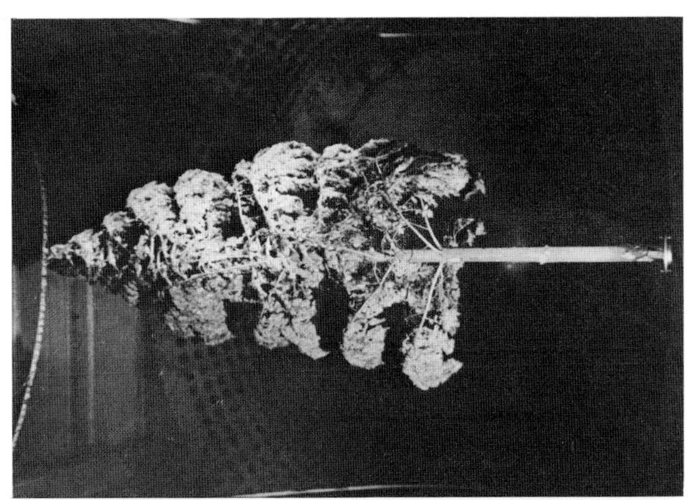

Plate 3 A specimen of Douglas Fir (*Pseudotsuga taxifolia*) mounted in a wind tunnel at the National Physical Laboratory. When the windspeed was increased from zero (left-hand) to 26 m s^{-1} (right-hand), there was a marked decrease in the area of foliage presented to the airstream and hence in the drag coefficient (p. 84). (Photographs by A. I. Fraser,[42] reproduced by permission of the Forestry Commission.)

Close to the earth's surface, turbulence is manifested by sudden, erratic and continuous changes of windspeed and direction, but, in restricted conditions that are difficult to achieve in practice, the *mean* windspeed increases logarithmically with height within a few metres of the surface. The main restrictions are:

(i) that the underlying surface is horizontal and uniform, i.e. composed of roughness elements (e.g. plants, trees) of more or less equal height and regular spacing;
(ii) that measurements are confined to the boundary layer characteristic of the underlying surface (p. 94);
(iii) that turbulence is generated exclusively by the mechanical forces of friction and form drag at the surface and not by buoyancy—air movements induced by vertical temperature gradients. In practice, this implies that the temperature gradient should not exceed a few hundredths of a degree Centigrade per metre;
(iv) that the windspeed is averaged over a period substantially longer than the characteristic period of fluctuations in the instantaneous windspeed. In practice, the averaging period should be at least 10 minutes and mean wind profiles are usually determined for intervals of 30 minutes or 1 hour.

Wind profiles

Figure 6.5 shows how windspeed increases with height over very short grass and over a tall farm crop when all these conditions are satisfied. In general, the windspeed $u(z)$ at height z increases linearly with $\ln(z-d)$ where z is the height measured above the ground surface and d is a datum level known as the **zero plane displacement**, determined experimentally. The value of d is usually between 0·6 and 0·8 of the height h of the roughness elements, i.e. the canopy height in the case of a farm crop, but the precise ratio of d to h depends on the spacing of the roughness elements and on the ratio of the accumulated area of each element to unit area of the underlying surface.

Stanhill[126] reviewed a large number of measurements of d for vegetation ranging in height h from 0·2 to 20 m and established a linear relation between $\log d$ and $\log h$:

$$\log d = 0{\cdot}9793 \log h - 0{\cdot}1536 \qquad 6.7$$

The determinations of d were too scattered to justify the quotation of constants so precisely and the much simpler relation $d = 0{\cdot}63h$ fits them as well as equation 6.7.

The relation between windspeed and height can be expressed formally by the equation

$$u(z) = a[\ln(z-d) - \ln z_0] \qquad 6.8$$

where z_0 is known as the **roughness length** and a is a parameter with the dimensions of velocity. The physical significance of a will emerge later and z_0 is determined in practice by extrapolating an observed linear relation between $u(z)$ and $\ln(z-d)$ to the point where $u=0$ at $z-d=z_0$ (Fig. 6.6). This extrapolation is simply a mathematical device to determine one of the constants in the wind profile equation. The plane

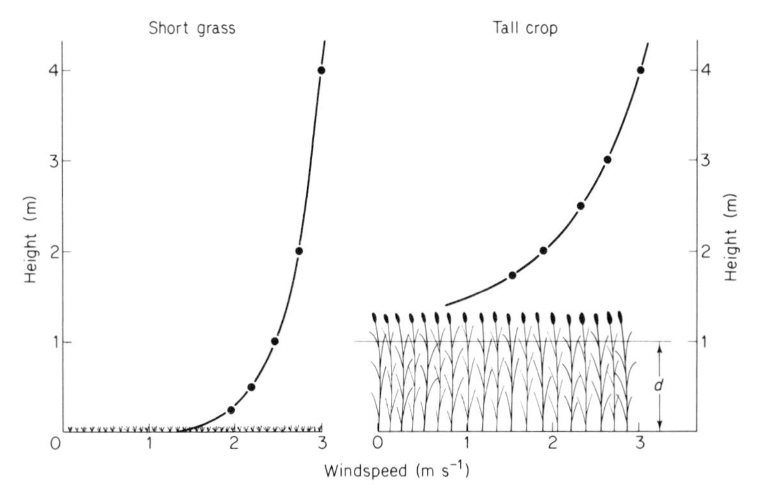

Fig. 6.5 Wind profiles over short grass and a tall crop when the windspeed at 4 m above the ground is 3 m s^{-1}. The filled circles represent hypothetical measurements from sets of anemometers.

$z=d+z_0$ can be regarded as an *apparent* sink of momentum in the canopy, but the linear relation between u and $\ln(z-d)$ is not valid below the top of the roughness elements and the real windspeed is finite at the height $d+z_0$. (The relation between windspeed and height within a crop canopy is discussed on p. 205). Differentiating equation 6.8 shows that the wind gradient $\partial u/\partial z$ is inversely proportional to $(z-d)$.

The roughness length of natural surfaces is about an order of magnitude less than the height of the roughness elements, e.g. 0·1 cm for a lawn ($h=1$ cm), 4 cm for barley ($h=70$ cm) and 20 cm for maize ($h=230$ cm). By correlating $\log z_0$ and $\log h$ for 12 species ranging in height from 2 cm to 6 m, Tanner and Pelton[136] established a relation

$$\log z_0 = 0{\cdot}997 \log h - 0{\cdot}883$$

which for all practical purposes is indistinguishable from $z_0 = 0{\cdot}13h$. The numerical factor in this expression should be regarded as valid for vegetation with an average density of foliage, say about 3 to 5 m^2 leaf surface per m^3 of canopy space. Lettau[67] suggested that the factor should increase as foliage density increases. His formulae allow z_0 to be estimated

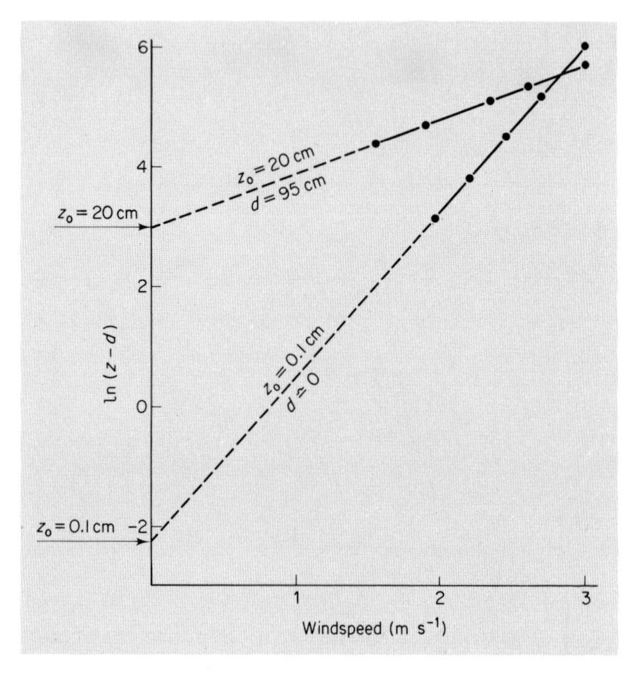

Fig. 6.6 Relations between windspeed and ln $(z-d)$ for the wind profiles in Fig. 6.5.

from the leaf area index of a crop projected in the plane of the wind or from the mean horizontal visual range inside a canopy.

The parameter a in equation 6.8 can be related to the shearing stress τ by introducing a transfer coefficient for momentum K_M analogous to the molecular transfer coefficient or kinematic viscosity. By analogy with equation 2.9a, the downward flux of momentum or the shearing stress at a height z in the boundary layer can be written in the form

$$\tau(z) = K_M \frac{\partial(\rho u)}{\partial z} \qquad 6.9$$

If the boundary layer is defined as a region in which $\tau(z)$ is constant with height and ρ is also constant, the product $K_M \, \partial(\rho u)/\partial z$ must be independent of z. But from equation 6.8, $\partial u/\partial z$ is proportional to $a/(z-d)$ so K_M must be proportional to $(z-d)$. This is a fundamental feature of momentum transfer in a turbulent boundary layer; subject to the conditions on p. 87, the coefficient of momentum transfer increases linearly with height above the zero plane d.

Now suppose $K_M = b(z-d)$ where b, like a, is a parameter with the dimensions of velocity. Then the equation of momentum transfer becomes $\tau = \rho \times (ab)$. The factor ab has the dimension of (velocity)^2 and is usually written u_*^2 where u_* is called the **friction velocity** because it is closely related to τ. Both a and b must be proportional to u_* and if $b = ku_*$ say, then $a = u_*/k$.

Substituting for a and b in previous equations gives three important relationships:

(i)
$$\tau = \rho u_*^2 \qquad 6.10$$

which defines the friction velocity as $u_* = (\tau/\rho)^{1/2}$

(ii)
$$u(z) = \frac{u_*}{k} \ln \frac{(z-d)}{z_0} \qquad 6.11$$

which is the conventional form of the wind profile equation. The factor k is known as von Karman's constant after a distinguished fluid dynamicist. The mean value of the most reliable field determinations is $k = 0{\cdot}41$.

(iii)
$$K_M = ku_*(z-d) \qquad 6.12$$

a relation which allows the turbulent transfer coefficient to be calculated from the wind profile using equation 6.11

As
$$\frac{\partial(u)}{\partial \ln (z-d)} = \frac{u_*}{k}$$

$$K_M = k^2(z-d)\{\partial u/\partial \ln (z-d)\} \qquad 6.13$$

and the right-hand side of this equation can be determined by plotting u against $\ln (z-d)$ once d is known. In practice, d is assumed to be $0{\cdot}7h$, or is determined by trial and error, or is calculated from three windspeeds u_1, u_2, u_3 measured at heights z, z_2, z_3. It can be shown from equation 6.8 that

$$\frac{u_1 - u_2}{u_1 - u_3} = \frac{\ln (z_1 - d) - \ln (z_2 - d)}{\ln (z_1 - d) - \ln (z_3 - d)} \qquad 6.14$$

The value of d can then be determined graphically from measurements of

$(u_1-u_2)/(u_1-u_3)$ if the right-hand side of equation 6.14 is plotted as a function of d.[115]

Equations 6.10 to 6.13 can now be used to determine the shearing stresses and transfer coefficients for the two wind profiles plotted in Fig. 6.6. The slope $\partial u/\partial \ln(z-d)$ is 36 cm s^{-1} for the short grass and 112 cm s^{-1} for the tall crop and the following table summarizes the differences between the two surfaces exposed to the same windspeed at an arbitrary height of 4 m.

	Short grass	Tall crop
Windspeed at m 4 (m s^{-1})	3	3
z_0 (cm)	0.1	20
d (cm)	0.7	95
u_* (ms^{-1})	0.15	0.46
τ (N m^{-2})	0.027	0.25
K_M at 4 m (m^2 s^{-1})	0.25	0.58

Note that the force of the wind on the rough surface is almost 10 times larger than on the smooth surface and that K_M for both surfaces is four orders of magnitude larger than the molecular transfer coefficient for momentum (0.15 cm^2 s^{-1}).

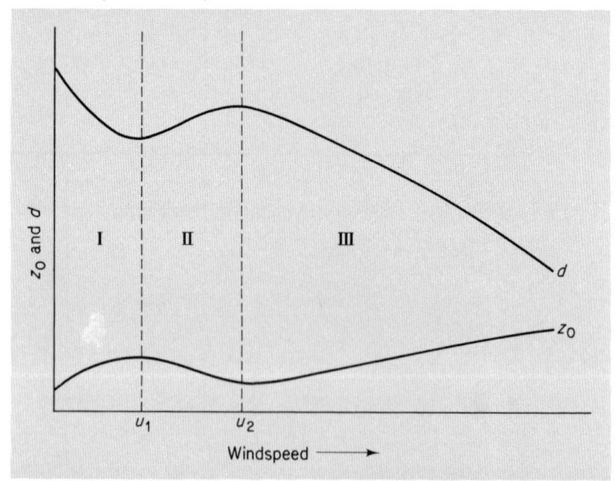

Fig. 6.7 Dependence of d and z_0 on windspeed, generalized from measurements on rice and other crops. The significance of the three regimes I, II, III is discussed in the text.

The behaviour of z_0 and d

Several workers have drawn attention to the way in which the values of z_0 and d appear to change systematically with windspeed when they are determined from wind profiles measured over a uniform stand of vegetation.[29, 85] Over some surfaces, for example, z_0 decreases with increasing windspeed and d is almost constant but over other surfaces z_0 increases with windspeed and d decreases. Figure 6.7 is an attempt to reconcile apparently conflicting accounts of the behaviour of z_0 and d. It is based on a series of measurements over rice crops exposed to a range of windspeeds from 0·5 to 10 m s^{-1} and it conceals a very large scatter in the original determinations.

In regime I, z_0 increases and d decreases as windspeed increases from

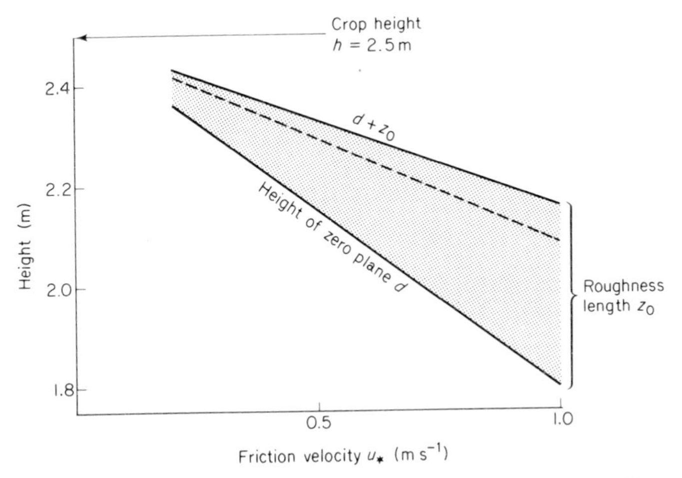

Fig. 6.8 Change of d and z_0 with u_* for maize as reported by Maki.[74] The line $d+z_0$ gives the height z at which the extrapolated windspeed is zero. The pecked line shows the estimated height $d+z_0$ when $z_0=k(h-d)$.

zero to a value u_1. This behaviour is not well documented for very low windspeeds but could be accounted for by a transfer of momentum from the surface of a canopy to deeper layers when the stems of plants begin to sway in a light wind. In regime II between u_1 and u_2, an increase of windspeed is accompanied by a decrease of z_0 and an increase of d. The change of z_0 may be a consequence of a movement of leaves into a more streamlined position (Plate 3) or of a decrease in drag coefficient as predicted by equation 6.6 when the Reynolds number for leaves approaches a value at which skin friction becomes unimportant in relation to form drag. When the windspeed exceeds u_2 (regime III), z_0 begins to increase again and d decreases. This behaviour can be ascribed to a substantial lowering of the canopy which can occur when plants bend in a light wind and to the very irregular way in which the canopy of a flexible crop can be deformed when turbulent eddies sweep over it.

This explanation for the behaviour of z_0 and d may be incorrect in detail but it draws attention to the importance of leaf size and of the flexibility of leaf petioles and plant stems in determining the aerodynamic behaviour of a crop stand. Regime II is well established for vegetation with relatively small, flexible leaves (grass, oats, barley) when the windspeed is less than 5 m s^{-1} and regime III appears to hold over a wide range of windspeeds for crops with larger and more rigid leaves[14] (wheat, maize, sorghum) (Fig. 6.8).

For a crop growing to a height of h, $h-d$ can be regarded as a measure

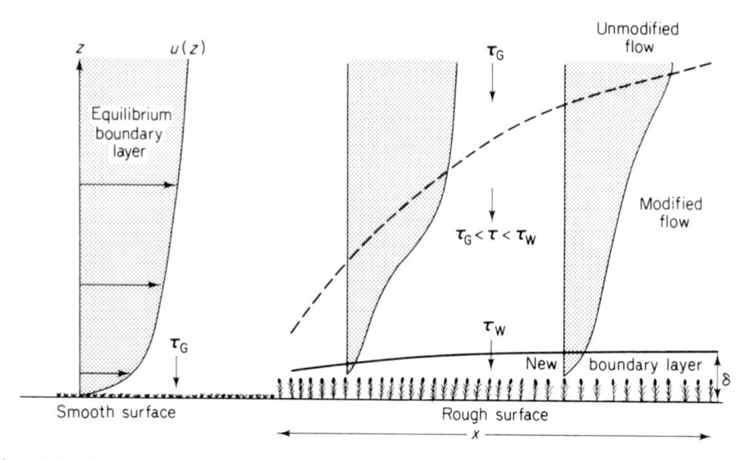

Fig. 6.9 Development of a new equilibrium boundary layer when air moves from a relatively smooth to a rougher surface. The ratio of the vertical to the horizontal scale is 20 :1. The broken line is the boundary between unmodified flow in which the vertical momentum flux is τ_G and modified flow in which the flux is between τ_G and τ_W. The flux is τ_W below the height δ.

of the depth of the layer in which momentum is absorbed. The complementary changes of z_0 and d shown in Fig. 6.8 are consistent with Thom's suggestion[138] that a first approximation to z_0 is given by $k(h-d)$ where k $(=0{\cdot}41)$ is von Karman's constant. This relation implies that z_0 is a measure of the average size of eddies at the top of the canopy. The representative values of $d=0{\cdot}63h$ and $z_0=0{\cdot}13h$ give $z_0=0{\cdot}35(h-d)$ which is only 15% less than $k(h-d)$. For rice, however, the ratio of $(h-d)$ to z_0 is not constant but appears to increase almost linearly with increasing windspeed.

Fetch

Just as the depth of a laminar boundary on a flat plate depends on the distance from the leading edge (p. 80), the depth of a turbulent boundary layer can be related to the **fetch** or distance of traverse across a uniformly rough surface. Figure 6.9 illustrates the cross section of a flat field of wheat adjacent to an equally flat playing field covered with short grass. The wheat will exert a substantially larger drag on the air than the grass so in appropriate wind directions the air will be decelerated as it moves from the smoother to the rougher surface. The experimental evidence from artificially roughened surfaces shows that the drag will change rapidly, say in 10 or 15 m, from τ_G on the grass to $\tau_W > \tau_G$ on the wheat. The transition in surface roughness affects the airflow to a height δ' and from measurements by Bradley[15] the relation between δ'(m) and the fetch x(m) was

$$\delta' = x^{0.8}/10 \qquad 6.15$$

during periods when the vertical temperature gradient was very small. On the basis of wind tunnel evidence that the flux of momentum is constant in the lowest 15% of the turbulent boundary layer, the logarithmic wind profile would be expected to extend to a height $\delta' = 0.015x^{0.8}$. For example, to get a constant flux layer 1 m deep, a fetch of about 200 m would be needed. In practice, many workers have had to be satisfied with fetch/height ratios much less than 200:1 partly because uniform fetches of several hundred metres are rare on experimental farms and partly because farm managers tend to be sceptical about rates of equilibration in the atmospheric boundary layer.

Other eddy transfer processes

In *laminar* boundary layers, momentum transfer is a molecular process determined partly by a velocity gradient and partly by a kinematic viscosity ν which depends on molecular agitation and hence on temperature. The corresponding processes of heat and mass transfer are a consequence of molecular agitation and are determined partly by gradients of temperature or concentration and partly by a coefficient similar to K_M (p. 91). In *turbulent* boundary layers the **eddy coefficient** of momentum transfer K_M depends on the rate of momentum transfer ($K_M \propto u_* \propto \sqrt{\tau}$) when the temperature gradient is small and increases linearly with height. It is helpful to think of turbulence as consisting of individual eddies playing the same role in turbulent transfer that individual molecules play in molecular transfer. The dependence of K_M on $z-d$ can be related to the fact that the size of eddies increases with distance from the surface and the dependence of K_M on u_* implies that eddies mix the atmosphere at a rate that increases with windspeed. The shape of the wind profile can be deduced from an initial assumption about the mean size of eddies and their relation to velocity fluctuation, the 'mixing length' theory of turbulence.

The analogy between molecules and eddies is limited in several important respects: in a closed system the number and size of molecules are fixed whereas eddies in turbulent flow are continually being generated, deformed, and destroyed so that the energy of eddy motion is ultimately degraded to heat, i.e. to the kinetic energy of individual molecules. The process by which eddies extract kinetic energy from the mean flow is fundamental to the maintenance of turbulence but the amounts of energy involved are so small in relation to radiation and convection that they are not relevant in this context.

Eddies are responsible for the transfer of heat, water vapour and carbon dioxide in the atmosphere when there is an appropriate potential gradient. Just as K_{M} is defined as the ratio of a momentum flux to the gradient $\partial(\rho u)/\partial z$, corresponding coefficients can be defined for heat (K_{H}), for water vapour (K_{V}) and for carbon dioxide (K_{C}).

Turbulent mixing plays an indispensable role in creating microclimates that can sustain life. Without turbulence in the boundary layer, the exchange of sensible heat between the atmosphere and the surface would involve enormous diurnal changes of temperature which plants could not endure. Furthermore, animals and humans would be asphyxiated by the products of their own metabolism.

Effects of buoyancy

The effects of frictional or mechanical turbulence are manifested by the swirling movement of fog, by the flapping of flags and by the discrete eddies that can be seen on a windy day weaving across a cornfield or whipping up spray from the sea surface near an exposed coastline. Another important mechanism for momentum transport is buoyancy, the vertical movement of parcels of air that are hotter or colder than their surroundings. The turbulence generated by buoyancy is made visible by the ascent of smoke above chimneys, by the formation of cumulus and by small scale fluctuations in the refractive index of heated air above a dry soil surface or above a roadway or fire.

When air temperature near the surface decreases rapidly with height, say at 1 °C per metre or more, the atmosphere is said to be 'unstable': any parcel of air pushed upwards, e.g. by turbulence or irregular topography, will tend to continue its ascent because it will be warm, light and therefore buoyant with respect to its surroundings. Conversely when temperature decreases with height, e.g. at 1°C per metre, the atmosphere is 'stable' because buoyancy forces oppose the original displacement. When the temperature gradient is very close to zero, −0·01°C per metre to be precise, a rising parcel of air would be cooled by expansion to exactly the same temperature as its surroundings and the air is then said to be in neutral equilibrium. A gradient of −0·01°C per metre is known as the dry adiabatic lapse rate and is usually given the symbol Γ.

In unstable air, turbulence is enhanced by buoyancy forces and in stable air it is suppressed. The effects of buoyancy on the shape of the wind profile and on rates of transfer by turbulence can be expressed by parameters that depend on the relation between the production of energy by buoyancy forces and the dissipation of energy by mechanical turbulence. The two best established parameters are the Richardson number Ri which can be calculated directly from gradients of temperature and windspeed and the Monin Obukhov length L, which is a function of the corresponding fluxes of heat and momentum.[155] In symbols

$$L = -\frac{\rho c_p T u_*^3}{kg\mathbf{C}} \tag{6.16}$$

where $\mathbf{C}$ = heat transfer by convection (positive upwards)
T = absolute temperature (K)
and g = gravitational acceleration

Webb[156] has shown that when z/L is between −0·03 and +1, i.e. from moderately unstable to very stable conditions, the transfer coefficient for momentum can be expressed as

$$K_M = ku_*z(1 + nz/L)^{-1} \tag{6.17a}$$

where n is a number to be determined empirically. The mean value of n in stable air was found to be 5·2, the mean value in unstable air was 4·5 and a round figure of 5 can be adopted in practice over the whole range of stability for which the equation was shown to apply. The limit $z/L = +1$ is an index of very stable air in which turbulence is completely suppressed and $z/L = -0{\cdot}03$ is a critical value below which heat transfer is dominated by free convection.

Other recent analyses of wind profile and shearing stress measurements suggest that over a wide range of unstable conditions K_M has the form

$$K_M = ku_*z(1 - 16z/L)^{1/4} \tag{6.17b}$$

which gives values similar to those of 6.17a with $n = 4$ when $16z/L$ is much less than unity.[36]

Resistance in the atmosphere

The equation for the increase of windspeed with height can be integrated to determine the diffusion resistance for momentum transfer between the level $z=d+z_0$, representing the momentum sink, and the atmosphere at a height z (p. 13). The resistance r_{aM} is defined by writing the flux of momentum in the form $\tau=\rho\{u(z)-u(d+z_0)\}/r_{aM}$. Although the *real* velocity at a height of $d+z_0$ is always finite, the *extrapolated*

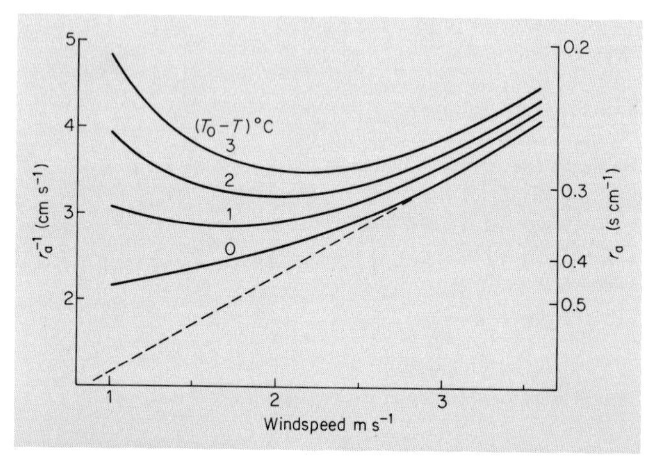

Fig. 6.10 Calculated values of resistance r_a in relation to windspeed and difference between surface and air temperature for a field of beans (*Vicia faba*). The roughness length decreased from 7 cm at a windspeed of 1 m s^{-1} to 3 cm at 3 m s^{-1}. Windspeed and temperature are referred to a height of 1·2 m above the zero plane.

velocity derived from equation 6·8 is always zero (because ln 1 = 0). Substituting $u(d+z_0)=0$ gives the effective resistance to momentum transfer in neutral conditions as

$$\begin{aligned} r_{aM} &= \rho u(z)/\tau = u(z)/u_*^2 \\ &= \{\ln (z-d)/z_0\}/(ku_*) \\ &= \{\ln (z-d)/z_0\}^2/(k^2u(z)) \end{aligned} \qquad 6.18$$

If z_0 were independent of windspeed, $1/r_{aM}$ would be proportional to $u(z)$ but for crops whose roughness length decreases with increasing windspeed, $1/r_{aM}$ is approximately constant over a range of low windspeeds (Fig. 6.10).

In non-neutral conditions, the resistance derived from equation 6.17 is

$$[ku_*]^{-1}\left[\ln\left(\frac{z-d}{z_0}\right)+\frac{n(z-d)}{L}\right]$$

provided $z-d$ is at least an order of magnitude greater than z_0. When $[\ln(z-d)/z_0)]/ku_*$ is identified as r_{aM}, the additional component $[n(z-d)/L]/ku_*$ can be regarded as a stability resistance which is positive in stable and negative in unstable conditions. The tendency for $1/r_{aM}$ to become independent of windspeed as the speed decreases is more pronounced in unstable than in neutral or stable conditions because the decrease in turbulent energy associated with decreased friction is compensated by an increase in the supply of energy from buoyancy (Fig. 6.10).

7

Heat Transfer—(i) Convection

All inflammable and combustible objects conceal in their substance, if nothing else, at least the particles which enable them to shoot out flames, radiate light, throw off sparks, and scatter glowing ashes far and wide.

Three mechanisms of heat transfer are important in the environment of plants and animals; *radiation*, governed by principles already considered in Chapter 2; *convection*, which is the transfer of heat by moving air; and *conduction* in solids and still gases which depends on the exchange of kinetic energy between molecules. Two types of convection are important in micrometeorology: 'forced' convection or transfer through the boundary layer of a surface exposed to an airstream, proceeding at a rate which depends on the velocity of the flow—a process analogous to skin friction; and 'free' convection which depends on the ascent of warm air above heated surfaces or the descent of cold air beneath cooled surfaces.

All these mechanisms of transfer are exploited in domestic heating systems. Fan heaters distribute hot air by forced convection; convector heaters and hot water 'radiators' circulate warm air by free convection; underfloor heating depends on the conduction of heat from cables buried below the floor; and the conventional bar-type radiator loses heat both by convection and radiation.

The analysis of convection is greatly simplified by using non-dimensional groups of quantities and a short description of these groups is needed to introduce a comparison of the convective heat loss from objects of different size and shape.

NON-DIMENSIONAL GROUPS

When a surface immersed in a fluid loses heat through a laminar boundary layer of uniform thickness δ, the heat transfer per unit area can be written as

$$\mathbf{C} = k(T_s - T)/\delta \qquad 7.1$$

where k is the thermal conductivity of the fluid, T_s is surface temperature and T is fluid temperature. The same equation can be used in a purely formal way to describe the heat loss by forced or free convection from any object with a mean surface temperature of T_s surrounded by fluid at T, even though the boundary layer is neither laminar nor uniformly thick. In this case, δ is the thickness of an equivalent rather than a real laminar layer. It is determined by the size and geometry of the surface and by the way in which fluid circulates over it. A more useful form of equation 7.1 can be derived by substituting a characteristic dimension of the body d for the equivalent boundary layer thickness δ which cannot be measured directly. For a sphere or cylinder, the diameter is a logical choice for d and for a rectangular plate d is the length in the direction of the wind. The equation then becomes

$$\mathbf{C} = \left(\frac{d}{\delta}\right) k(T_s - T)/d \qquad 7.2$$

The ratio d/δ is called the **Nusselt number** after its first exponent and is often written Nu. Just as the Reynolds number is a convenient way of comparing the forces associated with geometrically similar bodies immersed in a moving fluid, the Nusselt number provides a basis for comparing rates of convective heat loss from similar bodies of different scale exposed to different windspeeds.

The rate of convective heat transfer in air can be written as

$$\mathbf{C} = \rho c_p(T_s - T)/r_H \qquad 7.3$$

where r_H is a thermal diffusion resistance (p. 13). Comparison of equations 7.2 and 7.3 gives

$$r_H = \frac{\rho c_p d}{k\,\mathrm{Nu}} = \frac{d}{\kappa\,\mathrm{Nu}} \qquad 7.4$$

where κ is the thermal diffusivity of air.

Forced convection

Nusselt numbers can be expressed as a function of other non-dimensional groups. In forced convection, Nu is determined by heat transfer in the boundary layer over a surface that is hotter or cooler than the fluid passing over it and is therefore a function of the Reynolds number. In this case the dynamical similarity of different systems expressed by similar values of Re is closely related to the thermal similarity expressed by similar values of Nu. The value of Nu for forced convection is also a function of the **Prandtl number** $\mathrm{Pr}=\nu/\kappa$, a ratio related to the difference in boundary layer depth for momentum and heat transfer. Measurement of heat loss by forced convection from planes, cylinders and spheres can be described by the general relation

$$\mathrm{Nu} = \mathrm{Re}^n \, \mathrm{Pr}^m \qquad 7{\cdot}5$$

where n and m are numerical constants. In air, for which $\mathrm{Pr}=0{\cdot}71$ independent of temperature (p. 13), the Nusselt number for forced convection can be written simply as

$$\mathrm{Nu} = A \, \mathrm{Re}^n$$

Values of A and n for different types of geometry are given in Table A.5a (p. 224).

Free convection

In free convection, heat transfer depends on the circulation of fluid over and around an object, maintained by gradients of temperature which create gradients of density. In this case the Nusselt number is a function of another non-dimensional group, the **Grashof number** Gr as well as of the Prandtl number Pr. The Grashof number is determined by the temperature difference between the hot or cold object and the surrounding fluid $(T_s - T)$, the characteristic dimension of the object d, the coefficient of thermal expansion of the fluid ($a = 1/273$ for a perfect gas), the kinematic viscosity of the fluid ($\nu = 0{\cdot}15$ cm^2 s^{-1} for air) and the acceleration of gravity. Physically, the Grashof number is the ratio of a buoyancy force times an inertial force to the square of a viscous force. Numerically, it is calculated from

$$\mathrm{Gr} = agd^3(T_s - T)/\nu^2 \qquad 7{\cdot}6$$

In a system with a large Grashof number, free convection is vigorous because buoyancy and inertial forces which promote the circulation of air are much larger than the viscous forces which tend to inhibit circulation.

The Nusselt number for free convection in a gas is proportional to $(\mathrm{Gr}\,\mathrm{Pr})^n$ and can therefore be written as

$$\mathrm{Nu} = B \, \mathrm{Gr}^m \qquad 7{\cdot}7$$

for a specific gas such as air (Pr = 0·71). The numerical constants B and m which depend on geometry are tabulated in Appendix A.5b (p. 225). When appropriate values of a and ν for air at 20°C are inserted in equation 7.6, it can be shown that

$$\mathrm{Gr\,Pr} = 112\, d^3(T_s - T) \tag{7.6a}$$

and

$$\mathrm{Gr} = 158\, d^3(T_s - T) \tag{7.6b}$$

where d is the characteristic dimension in centimetres.

Note that for laminar free convection, $m = 1/4$ irrespective of the shape of the object losing heat. In this case, the rate of convective heat loss is proportional to $(T_s - T)^{5/4}$, the so-called five-fourths power law of cooling.

Criteria for forced and free convection

It is not always obvious *a priori* whether the heat transfer from an object is likely to depend on forced or on free convection. As a rough criterion for distinguishing the two regimes, the Grashof number may be compared with the square of the Reynolds number. As Gr depends on $\frac{\text{buoyancy} \times \text{inertial forces}}{(\text{viscous forces})^2}$ and Re^2 depends on (inertial forces)^2/(viscous forces)^2, Gr/Re^2 is proportional to the ratio of buoyancy to inertial forces. When Gr is much larger than Re^2, buoyancy forces are much larger than inertial forces and heat transfer is governed by free convection. When Gr is much less than Re^2, buoyancy forces are negligible and forced convection is the dominant mode of heat transfer.[61]

As rules of thumb derived from limited experimental evidence, the appropriate function of the Grashof number should be used to calculate convective heat loss in cases where $\mathrm{Gr} > 16\,\mathrm{Re}^2$ and the appropriate function of the Reynolds number should be used when $\mathrm{Gr} < 0{\cdot}1\,\mathrm{Re}^2$. For intermediate values of Gr/Re^2, Nu should be calculated both for forced and for free convection and the larger number should be used to estimate the rate of heat transfer.

For example, when a leaf with $d = 5$ cm is 5°C warmer than the surrounding air its Grashof number is about 10^5 whereas Re^2 is about $10^7\,V^2$ when V is in m s^{-1}. A regime of forced convection is expected when V exceeds 1 m s^{-1} but at windspeeds between 0·1 and 0·5 m s^{-1}, which are often found in crop canopies, both forced and free convection will be active mechanisms of heat transfer.

A cow with $d = 0{\cdot}5$ m and a surface temperature 20°C above the ambient air has $\mathrm{Gr} = 4 \times 10^8$ and $\mathrm{Re}^2 = V^2 \times 10^9$ when V is in m s^{-1}. In this case, free convection will be the dominant form of heat transfer when the animal is exposed to a light draught indoors but at windspeeds of the order 1 m s^{-1} in the field, the convection regime will again be mixed.

Criteria for laminar or turbulent flow

Both in forced and in free convection, the size of the Nusselt number depends on whether the boundary layer is laminar or turbulent. In an airstream free from turbulence, the transition from laminar to turbulent flow in the boundary occurs at Reynolds number of the order of 10^5 but in a turbulent airstream the critical Reynolds number decreases to an extent that depends partly on the amplitude of the velocity fluctuations and partly on their frequency. In micrometeorological problems involving leaves or other plant organs, Re is usually between 10^3 and 10^4 but it has never been clearly demonstrated whether the boundary layer of a leaf in a crop canopy, for example, should be regarded as laminar or turbulent. At a Reynolds number of 10^4, the Nusselt number for laminar forced convection from a flat plate is $0{\cdot}60 \times (10^4)^{1/2}$ or 600 compared with $0{\cdot}032 \times (10^4)^{0{\cdot}8}$ or 510 for a turbulent boundary layer; and at $\text{Re} = 4 \times 10^4$, the corresponding numbers are 1200 at 1500. Thus for values of Re in the range of micrometeorological interest, there will usually be little difference between the conventional Nusselt numbers for laminar and turbulent boundary layers. It does not necessarily follow that the same Nusselt numbers will be valid when the airstream itself is turbulent. Parlange, Waggoner and Heichel[100] claim that the heat loss from a 20 × 20 cm section of tobacco leaf exposed to a turbulent airstream was about 2·5 times larger than the loss predicted from the Nusselt number for laminar flow.

The onset of turbulence in free convection occurs when the Grashof number exceeds 10^8, an unusual situation in micrometeorology. For example, the surface temperature of a sheep or a man would need to be at least 30°C above the temperature of the ambient air to achieve $\text{Gr} = 10^8$. The assumption of laminar flow will therefore be valid in cases of free convection as well as in forced convection.

MEASUREMENTS OF CONVECTION

Plane surfaces

When the boundary layer over a plane surface is laminar, the rate of heat transfer between the surface and the airstream can be calculated from first principles for two discrete cases. First, if the temperature is uniform over the whole surface the Nusselt number is

$$\text{Nu} = 0{\cdot}66\ \text{Re}^{0{\cdot}5}\text{Pr}^{1/3} \qquad 7.8$$

and this relation is quoted in engineering texts which are concerned mainly with the heat transfer from metal surfaces with high thermal conductivity.

In the second case which is more biologically relevant, the heat flux per unit area is constant over the whole surface. This condition should be valid for poor thermal conductors exposed to a uniform flux of radiation, e.g. leaf laminae in sunshine. The uniformity of heat flux from a leaf surface has not been established experimentally but it is clear from thermocouple measurements of leaf temperature in natural conditions (Fig. 7.1)

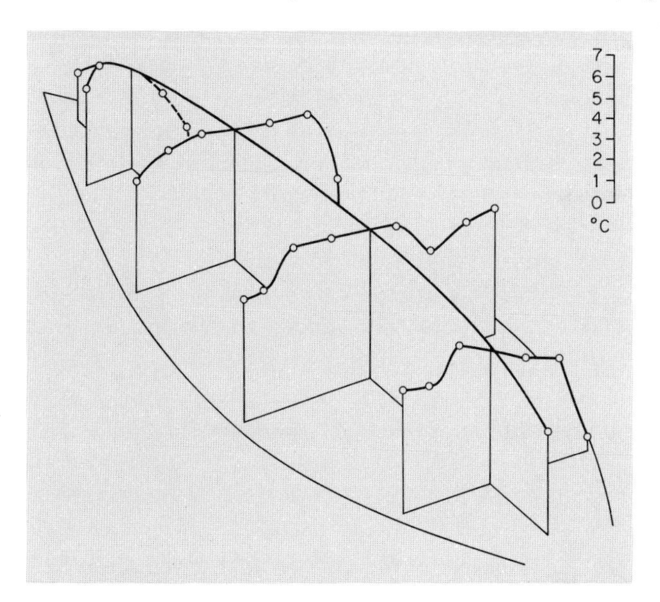

Fig. 7.1 Temperature distribution over a leaf of *Canna indica*. Mean excess temperature 5·5°C, windspeed 3·3 m s^{-1}, net radiation 680 W m^{-2}, leaf dimensions 34 × 14·5 cm (from Raschke[113]).

that it is not legitimate to treat sunlit leaves as isothermal surfaces. According to Parlange *et al.*,[100] the assumption of a uniform heat flux leads to the prediction that the excess leaf temperature $T_s - T$ should increase with the square root of the distance from the leading edge x (cf. the uniform temperature case in which the *flux* decreases with the square root of x). It is convenient to incorporate x in a local Reynolds number $\mathrm{Re}_x = (Vx/\nu)$. In laminar flow the excess temperature then becomes

$$T_s - T = 2{\cdot}21\ \mathbf{C}\frac{k}{x}(\mathrm{Re})^{-1/2}(\mathrm{Pr})^{-1/3}$$

and the mean temperature over a plate of length d is

$$\frac{\int_0^d (T_s - T)\,dx}{\int_0^d dx} = 1{\cdot}47\,\mathbf{C}\frac{k}{d}\,\mathrm{Re}^{-1/2}\,\mathrm{Pr}^{-1/3}$$

The mean Nusselt number defined as $\overline{\mathrm{Nu}} = \mathbf{C}d/k\bar{\theta}$ is

$$\overline{\mathrm{Nu}} = 0{\cdot}68\,\mathrm{Re}^{1/2}\,\mathrm{Pr}^{1/3}$$

which is only a few per cent larger than the Nusselt number for the uniform temperature case. (Other workers have given an erroneous value of 0·90 for the constant in this equation.)

Leaves

The heat loss from a model leaf was studied by sandwiching a length of resistance wire between two thin sheets of aluminium cut to the shape shown in Fig. 6.3.[137] The temperature distribution was nearly uniform over the surface of the lamina and when the flow was parallel to the leaf ($\phi = 0$) the Nusselt number for heat transfer from one side of the leaf was

$$\mathrm{Nu} = 0{\cdot}70\,\mathrm{Re}^{1/2}\,\mathrm{Pr}^{1/3} \qquad 7{\cdot}9$$

This is only a few per cent greater than the value of Nu for a *wide* plate of *uniform* length (Table A.5). The corresponding resistance to heat transfer is $d/(\kappa\mathrm{Nu})$ and, by adjustment of terms in equation 7.9, it can be shown that

$$r_H = \frac{1{\cdot}42}{u}\,\mathrm{Re}^{1/2}\,\mathrm{Pr}^{2/3}$$

With $\phi = 0$, the resistance of the leaf to momentum transfer by skin friction was

$$r_M = \frac{1{\cdot}42}{u}\,\mathrm{Re}^{1/2}$$

so the ratio of the two resistances is

$$\frac{r_H}{r_M} = \mathrm{Pr}^{2/3} = \left(\frac{\nu}{\kappa}\right)^{2/3} = 0{\cdot}83 \qquad 7.10$$

This simple relation, valid only when form drag is very much smaller than skin friction, demonstrates the analogy between heat and momentum transfer in laminar boundary layers. The factor $\mathrm{Pr}^{2/3}$ shows how the ratio of resistances depends on the relative boundary layer thickness for the two processes. The effective boundary layer thickness δ for the molecular transfer of an entity with a diffusion coefficient of X is proportional to

$X^{1/3}$, and the corresponding resistance δ/X is proportional to $X^{-2/3}$ as shown by equation 7.10.

A similar relation between resistance and windspeed was derived by Impens[54] who simulated natural leaves with pieces of blotting paper suspended in a stand of *Phaseolus vulgaris*. The heat transfer from *real* leaves of *Canna indica* was studied by Raschke[113] whose results can be expressed in the form

$$\mathrm{Nu} = 1{\cdot}5\ \mathrm{Re}^{1/2}\ \mathrm{Pr}^{1/3} \qquad 7.11$$

more than twice the accepted value for flat plates at a uniform temperature and close to the value reported by Parlange *et al.* Raschke showed that there was a substantial gradient of temperature across the leaf (Fig. 7.1) consistent with more effective heat exchange at the edges than at the centre. The remaining difference between equations 7.9 and 7.11 can be ascribed to turbulence in the airstream, surface roughness and free convection. For a leaf with $d = 15$ cm and an excess temperature of 10°C, the Grashof number (4×10^6) is of the same order as Re^2 at a windspeed of 2 m s^{-1} so the convection regime in Raschke's system was probably mixed.

The Nusselt number for leaves smaller than those used by Raschke is likely to fall between $0{\cdot}6\ \mathrm{Re}^{1/2}$ and $1.4\ \mathrm{Re}^{1/2}$ (putting $\mathrm{Pr}^{1/3} = 0{\cdot}9$). It is very difficult to account for the much larger values that have been deduced from analyses of the heat balance of leaves in a corn crop where the surface temperature was measured with a radiometer.[51]

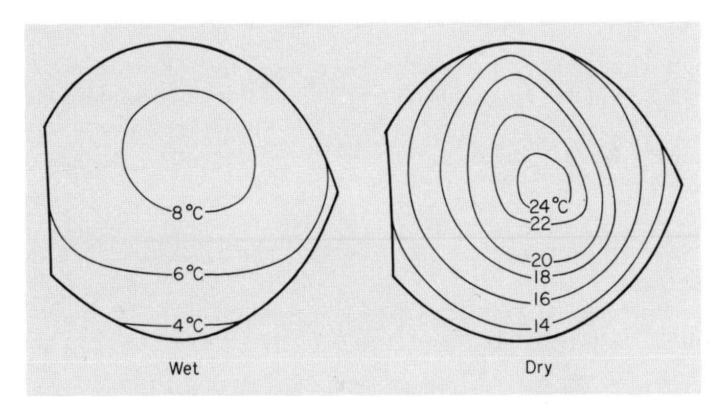

Fig. 7.2 Temperature distribution over the cardboard replica of a bean (*Phaseolus*) leaf held vertically in still air and exposed to a net radiant flux of 100 W m^{-2}. The isotherms of leaf minus air temperature were drawn from measurements with a radiometer at 21 points over the surface. Note that temperature gradients were substantial when the leaf was wet although much larger gradients were observed with a dry leaf (from Simmons[125]).

Little work has been published on pure *free* convection from flat leaves despite its significance in dense vegetation where the windspeed is often less than 0·5 m s^{-1}. Figure 7.2 shows a striking example of the excess temperatures measured on a vertical sheet of dry cardboard cut to the shape of a *Phaseolus* leaf.[125] The 'leaf' was exposed to a net radiant flux of 100 W m^{-2}, comparable with the flux that might be absorbed by a vertical leaf in bright sunshine. The vertical gradient of temperature shows how a skin of air in contact with the surface is progressively heated as it rises from the bottom to the top of the leaf. When the cardboard was saturated with water to simulate a transpiring leaf, the temperature excess was greatly reduced.

The dissipation of heat from a set of copper plates exposed to low windspeed was studied by Vogel.[152] All the plates had the same area but their shapes ranged from a circle and a regular 6-point star to replicas of oak leaves with characteristic lobes (Fig. 7.3). The amount of electrical

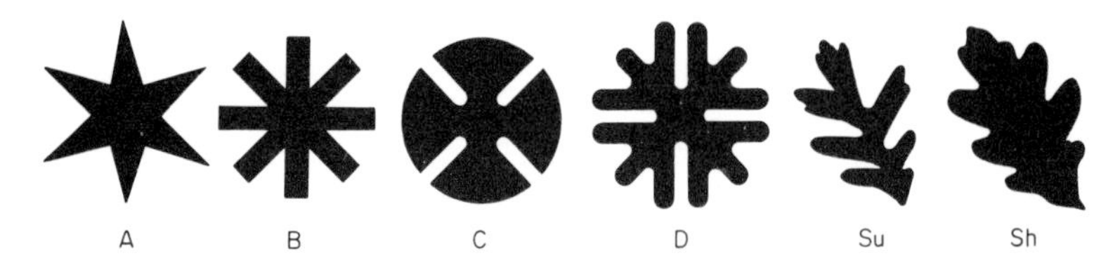

Fig. 7.3 Metal replicas of leaves used by Vogel[152] to study heat losses in free convection. The shapes Su and Sh represent sun and shade leaves of white oak.

energy needed to keep each plate 15°C warmer than the surrounding air was recorded at windspeeds from 0 to 0·3 m s^{-1} and at different orientations. This energy is proportional to Nu and inversely proportional to the resistance r_H under the conditions of the experiment. The main conclusions were:

(i) increasing airflow from 0 to 0·3 m s^{-1} decreased by r_H by 30 to 95%: the decrease was greater for the leaf models than for the stellate shapes;

(ii) the resistances of all the stellate and lobed plates were smaller than the resistance of the round plate and were less sensitive to orientation;

(iii) a deeply lobed model simulating a sun leaf of oak always had a smaller resistance than a shade leaf with smaller lobes;

(iv) the resistance of the leaf models was least when the surface was oblique to the airstream;

(v) serrations about 5 mm deep on the periphery of the circular plate had no perceptible effect on its thermal resistance;

(vi) the measurements could not be correlated using a simple Nusselt number based on a weighted mean width as described below.

These measurements support the hypothesis favoured by some ecologists that the shapes of leaves may represent an adaptation to their thermal environment. Because the natural environment is very variable and because physiological responses to changes of leaf tissue temperatures are complex, conclusive experimental proof is still lacking.

A Nusselt number for serrated or compound leaves can be calculated from an appropriate average length in the direction of the airstream.

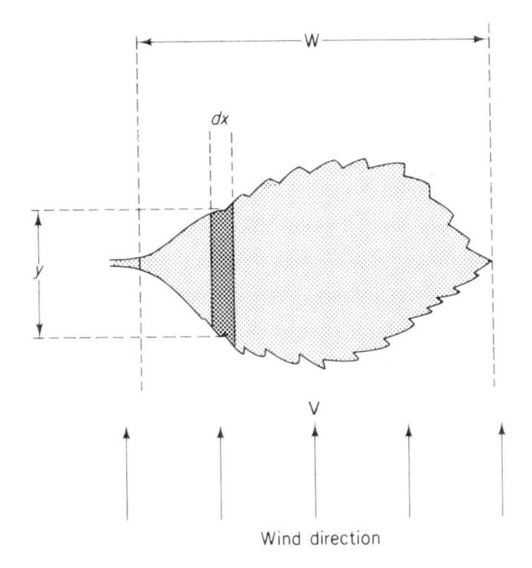

Fig. 7.4 Coordinates for integrating heat loss over the surface of a leaf of irregular shape.

If W is the width of a leaf at right angles to the flow, the leaf area can be expressed as $\int_0^w y\,dx$ (Fig. 7.4). When the Nusselt number is $A\,\mathrm{Re}^n$, an effective mean length $\bar{y}$ can be defined by writing the total heat loss from the leaf as

$$\mathbf{C} = A\left(\frac{V\bar{y}}{\nu}\right)^n \frac{k}{\bar{y}}\Delta T \times \int_0^w y\,dx \qquad 7.12$$

But the total heat loss can also be written in the form

$$\mathbf{C} = \int_0^w A\left(\frac{Vy}{\nu}\right)^n k\,\Delta T\,dx \qquad 7.13$$

and by equating these expressions the mean length is given by

$$\bar{y} = \left\{ \frac{\int_0^w y^n \, dx}{\int_0^w y dx} \right\}^{1/(n-1)} \tag{7.14}$$

For laminar forced convection, $n = 0{\cdot}5$. Equation 7.14 is also valid for free convection with $n = 0{\cdot}75$. Parkhurst *et al.*[99] measured the heat loss and excess temperature from a series of metal leaf replicas with a wide range of shapes. Almost all the Nusselt numbers based on the mean dimension $\bar{y}$ lay between the values $0{\cdot}60 \text{ Re}^{1/2}$ and $0{\cdot}80 \text{ Re}^{1/2}$.

Other work on the heat transfer of leaves includes estimates of r_H from the rate at which leaves cool when they are shaded; determinations of Nu in turbulent as well as in laminar conditions (Nu becomes proportional to $V^{0{\cdot}8}$ instead of $V^{0{\cdot}5}$); and an investigation of the effect of hairs, discussed on page 115.

Cylinders and spheres

The flow of air over cylinders is more complex than the flow over flat plates because of the separation of the boundary layer at the rear of the cylinder and the formation of vortices in the slipstream. The Nusselt number can be related to the Reynolds number by writing $\text{Nu} = A \text{ Re}^n \text{ Pr}^{1/3}$ but, unlike the corresponding quantities for flat plates, both A and n are functions of Re (Appendix A.5).

Mammals

In nature, many mammals and human beings can be regarded as horizontal or vertical cylinders for the purposes of estimating their convective heat loss. To establish convection coefficients, the heat exchange of animals is usually determined in calorimeters or climate rooms, but as these enclosures are seldom ventilated by a uniform horizontal airstream, only a few sets of measurements can be manipulated to determine the relationship between Nusselt and Reynolds numbers. For example, Joyce[56] and his colleagues found that the external conductance for a sheep was $(0{\cdot}15 + 0{\cdot}01 \, V^{0{\cdot}5})$ in units of $\text{M cal m}^{-2} \, 24 \, \text{h}^{-1} \, {}^\circ\text{C}^{-1}$ (see p. 121). The first term in this expression represents the contribution from radiative exchange and the second term is the convective component. The Nusselt number calculated from this component is plotted against the corresponding Reynolds number in Fig. 7.5. (Logarithmic scales are often used in diagrams of this type so that a relation of the form $\text{Nu} = A \text{ Re}^n$ becomes $\log \text{Nu} = n \log \text{Re} + \log A$. The quantities n and A can then be found from the slope and intercept respectively.) The Nusselt numbers for the sheep were close to the values calculated for a cylinder of 30 cm diameter using the standard relation $\mathrm{Nu}=0{\cdot}24\ \mathrm{Re}^{0\cdot60}$ for the appropriate range of Reynolds numbers. The relation $\mathrm{Nu}=0{\cdot}65\ \mathrm{Re}^{0\cdot5}$ fits the measurements even better.

Rapp[112] reviewed several sets of measurements on the heat loss from nude subjects exposed in different postures to a range of environments and found excellent agreement between measured and predicted Nusselt numbers except when the convection regime was mixed. For example, in one set of measurements the heat loss from vertical subjects standing in a horizontal airstream was $8{\cdot}5V^{0\cdot5}$ W m^{-2} (V in m s^{-1}) and for a man

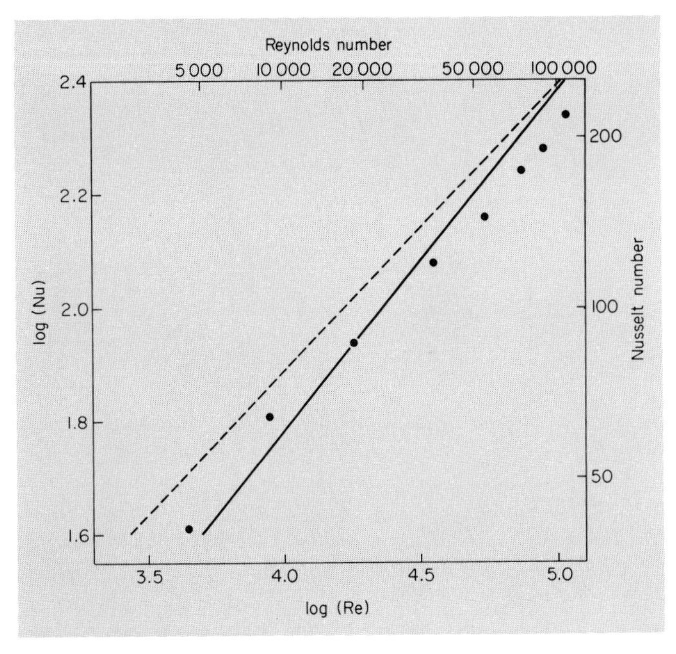

Fig. 7.5 Relation between log Nu and log Re. The continuous line is a standard relation for cylinders at right angles to the airstream, i.e. $\mathrm{Nu}=0{\cdot}24\ \mathrm{Re}^{0\cdot60}$. The filled points were calculated from measurements on sheep and the pecked line from measurements on human subjects.

with a characteristic dimension of 33 cm this is equivalent to a Nusselt number of $0{\cdot}78\ \mathrm{Re}^{0\cdot5}$, a relation shown by the pecked line in Fig. 7.5. Over a wide range of Reynolds numbers from $2{\cdot}3\times10^3$ to 10^5, the Nusselt number for the man is about 20% greater than the number for a sheep. This figure illustrates the usefulness of non-dimensional groups in comparing measurements of heat loss from different series of experiments.

In a light wind, the heat loss from sheep and other animals may be governed by free convection, particularly when there is a large difference of temperature between the coat and the surrounding air. When Merinos were exposed to strong sunshine in Australia, fleece tip temperatures reached 85°C when the air temperature was 45°C.[73] With $d = 30$ cm, the corresponding Grashof number is 2×10^8, so free and forced convection would be of comparable importance when $\mathrm{Re}^2 = 2 \times 10^8$, i.e. when windspeed was about 0·7 m s^{-1}. To calculate fleece temperatures in similar conditions, Priestley[110] used a graphical method to allow for the transition from free to forced convection with increasing windspeed (see p. 103).

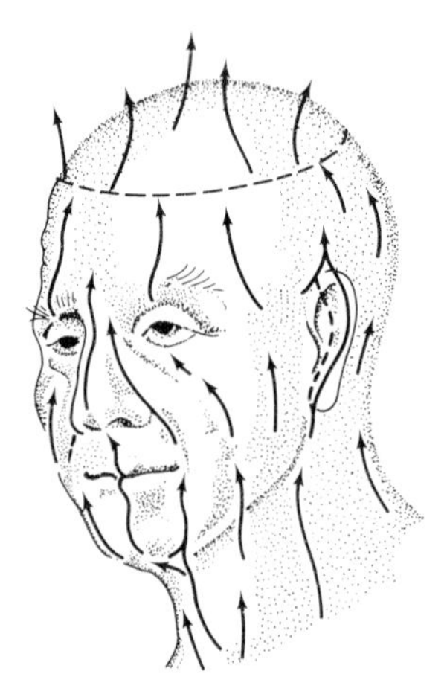

Fig. 7.6 Routes taken by the flow of naturally convected air over the human head (from Schlieren photography by Lewis *et al.*[68]).

For nude human subjects, free convection will be the dominant mode of convective heat loss at velocities up to 0·5 m s^{-1} provided the skin is at least 10°C warmer than the surrounding air. The movement of air associated with free convection from the head and limbs has been demonstrated by Lewis[68] and his colleagues using the technique of Schlieren photography (Fig. 7.6, Plate 4). The way in which the air ascending over the face is deflected from the nostrils may be important in preventing the inhalation of bacteria and other pathogens. Figure 7.7 shows that the velocity and temperature profiles close to a bare leg are characteristic of the flow in free convection from vertical surfaces.

The application of conventional heat transfer analysis is much more difficult for non-cooperative subjects like piglets, particularly when the

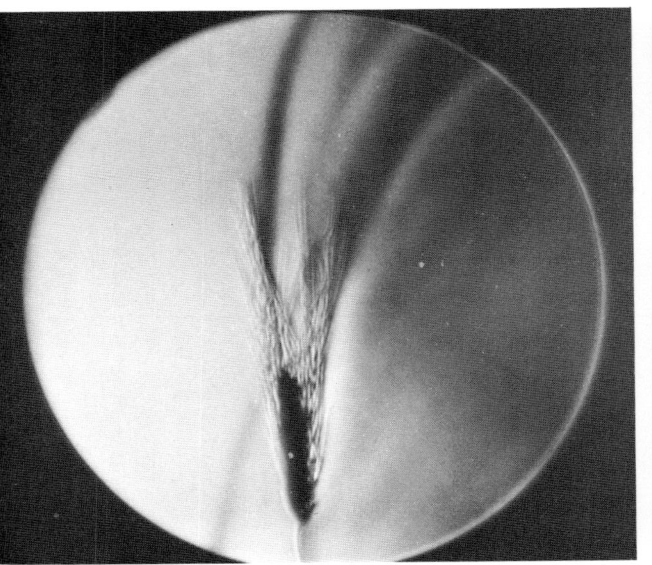

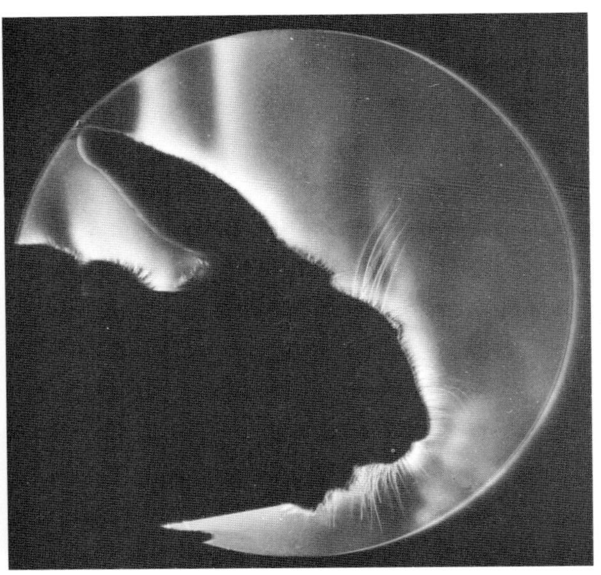

Plate 4 Schlieren photographs of an ear of awned wheat and a rabbit's head showing regions of relatively warm air (light) and cooler air (dark). The wheat was irradiated with a source giving about 800 W m^{-2} net radiation. Note the disturbed air round the rabbit's nostrils, the separation of rising air over the eyebrows, and the evidence of strong heating of air round the ears (see p. 112).

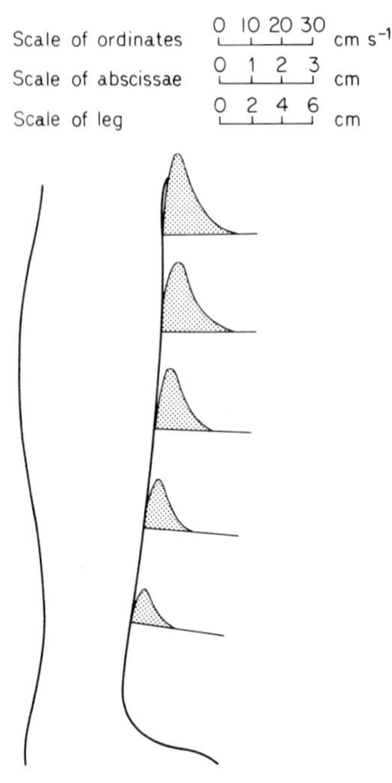

Fig. 7.7 Development of velocity profiles on the front of the leg measured with a hot wire anemometer (from Lewis *et al.*[68]).

animal changes its posture and orientation with respect to windspeed in response to heat or cold stress.

Insects

The convective heat loss from insect species was measured by Digby[32] who mounted dead specimens in a wind tunnel with transparent walls. Heating was provided by radiation from an external 2 kW lamp. The difference between the temperature of the insects and the ambient air was (i) directly proportional to incident radiant energy and (ii) inversely proportional to the square root of the windspeed as predicted for cylinders in the appropriate range of Reynolds numbers.

Evidence for the variation of excess temperature with body size was more difficult to interpret. When the maximum breadth of the thorax was taken as an index of body size d, the excess temperature was expected to change with a power of d between 0·5 (cylinders) and 0·4 (spheres). For the locusts *Schistocerca gregaria* and *Carausius morosus*, the exponent of d was 0·4 but for 20 species of Diptera and Hymenoptera, Digby's measurements support an exponent of unity, i.e. excess temperature was directly proportional to the linear dimensions of the insect. This anomalous result may be a consequence of considerable scatter in the measurements or of changes in the distribution of absorbed radiant energy with body size.

In a similar series of measurements, Church[24] used radiation from a 20 kHz generator to heat specimens of bees and moths. He argued that the amount of energy produced by an insect in flight would be roughly proportional to body weight and therefore used radiant flux densities that were proportional to the cube of the thorax diameter. When the insects were shaved to remove hair, the excess temperature was proportional to $d^{1\cdot4}$. As the heat loss per unit area was proportional to d, the Nusselt number was proportional to $d^{0\cdot6}$, close to expectation for a sphere. The excess temperature of insects covered with hair was about twice the excess for denuded insects showing that the thermal resistances of the hair layer and the boundary layer were of similar size.

If the proportionality between Nu and $d^{0\cdot6}$ is consistent with $\mathrm{Nu}=A\,\mathrm{Re}^n$, the index n must be 0·6 and Nu should be proportional to $u^{0\cdot6}$. In fact, Church found that the excess temperature of a shaved *Bombus* specimen was proportional to $u^{0\cdot4}$ but he explained this discrepancy in terms of the difference between temperatures at the surface of and within the thorax. If all the experimental results are assumed consistent with $n=0\cdot5$, the measurements on the *Bombus* specimen give a Nusselt number of $0\cdot28\,\mathrm{Re}^{0\cdot6}$, close to the standard relation for a sphere, $\mathrm{Nu}=0\cdot34\,\mathrm{Re}^{0\cdot6}$. The values of Nu from this equation are close to the values predicted from either of the relationships in Table A.5 within the appropriate range of Reynolds numbers.

Leaves

Many coniferous trees have needle-like leaves that can be treated as cylinders in order to estimate heat losses, though their cross section is seldom circular. Because the needles are closely spaced they shield each other from the wind and do not behave like isolated cylinders. Engineers have developed empirical formulae for the heat loss from banks of cylinders in regular arrays and their relevance to coniferous branches has been examined by Tibbals and his colleagues.[140] To avoid the difficulty of measuring the surface temperature of real needles, branches of pine, spruce and fir were invested with dental compound and then burnt to leave a void that was filled with molten silver. The compound was then removed leaving a silver replica of the branch—a modern version of the Midas touch! For Blue spruce (*Picea pungens*), the Nusselt numbers based on the diameter of single needles were about one third to one half of Nu for isolated cylinders at the same Reynolds number, a measure of mutual sheltering. For White fir (*Abies concolor*), the Nusselt number in transverse flow (across four rows of needles) was about 60% greater than in longitudinal flow (across 20 or 30 rows of needles). When an average of the two modes was taken, Nu was close to the experimental values for banks of tubes and similar agreement was obtained for *Pinus ponderosa*. In still air, the Nusselt number for all three species was close to an experimental value for horizontal cylinders at the same Grashof number.[99]

TRANSFER IN THE ATMOSPHERIC BOUNDARY LAYER

The earth's surface is rarely in thermal equilibrium with the air flowing over it and turbulence within the boundary layer is an effective mechanism for heat transfer between the surface and the free atmosphere. The process of heat transfer by turbulence is analogous to the process of momentum transfer discussed in Chapter 6 and is described by a similar equation relating the heat flux to a temperature gradient. If **C** is the heat transfer by convection and K_H is the turbulent transfer coefficient for heat

$$\mathbf{C} = -K_H \frac{\partial(\rho c_p T)}{\partial z} \tag{7.15}$$

The relation between K_H and the corresponding quantity for momentum K_M has been one of the most controversial topics of micrometeorology for at least 30 years. It has been argued that when the air near the ground becomes more unstable as a result of an increasing lapse rate, K_H should increase faster than K_M because of a preferential upward transport of heat in parcels of warm rising air. Conversely K_H is expected to be less than K_M in stable conditions when buoyancy tends to suppress heat transfer. Recent papers support conflicting conclusions which can be summarized as follows:

(i) $K_H > K_M$ in all unstable conditions $(z/L < 0)$[36, 101] or

(ii) $K_H = K_M$ in stable and in *moderately* unstable conditions[155, 156] $(-0{\cdot}03 < z/L < 1)$

(L is the Monin Obukhov stability length introduced on p. 97 and z is height.) When z/L is less than $-0{\cdot}05$, free convection becomes the dominant mechanism sustaining heat loss from the surface and there is general agreement that K_H exceeds K_M in this regime.

A method of calculating **C** from temperature and wind profiles will now be considered and to simplify the analysis K_H and K_M will be assumed equal. Within a few metres of the ground, ρ can be assumed constant. The identity of the eddy coefficients then implies that

$$-\frac{\rho c_p\, \partial T/\partial z}{\mathbf{C}} = \frac{\rho\, \partial u/\partial z}{\tau} \tag{7.16}$$

Within the boundary layer characteristic of the surface, $\mathbf{C}$ and $\boldsymbol{\tau}$ are independent of height by definition (p. 87) and when this condition is satisfied, equation 7.16 implies that T and u must be the same function of height. Thus when T and u are measured at a series of heights above a uniform surface, a graph of T plotted against u should be a straight line with slope $(\partial T/\partial u)$.

By rearranging equation 7.16 and setting $\boldsymbol{\tau}$ equal to ρu_*^2, it can be shown that

$$\begin{aligned} \mathbf{C} &= c_p(\partial T/\partial u)\boldsymbol{\tau} \\ &= \rho c_p(\partial T/\partial u)u_*^2 \end{aligned} \qquad 7.17$$

This is a convenient equation for determining the convective heat flux from measurements of windspeed and temperature at a series of heights within the boundary layer. The gradient $\partial T/\partial u$ is found graphically and in conditions near to neutral stability, the friction velocity u_*^2 can be found by plotting u against $\ln(z-d)$ (Fig. 6.6). In non-neutral stability the analysis is slightly more complicated because the slope of the wind profile is determined by the value of $\mathbf{C}$ as well as by u_*. The simplest procedure is to define the Monin Obukhov length as

$$\begin{aligned} L &= \frac{-u_*^3 T\rho c_p}{kg\mathbf{C}} \\ &= \frac{-u_* T}{kg(\partial T/\partial u)} \end{aligned} \qquad 7.18$$

from equation 7.17. Then the wind profile equation (p. 91) can be written in the form

$$u = \frac{u_*}{k}\left\{\ln\left(\frac{z-d}{z_0}\right) + \frac{n(z-d)kg(\partial T/\partial u)}{u_* T}\right\} \qquad 7.19$$

or

$$u - n(z-d)\frac{\partial T}{\partial u}\frac{g}{T} = \frac{u_*}{k}\ln\left(\frac{z-d}{z_0}\right)$$

The friction velocity u_* can now be determined by regarding

$$u - n(z-d)\frac{\partial T}{\partial u}\frac{g}{T}$$

as a form of corrected velocity which can be plotted against $\ln(z-d)$ to obtain u_*/k as a slope (see p. 89). The convective heat flux is then calculated by inserting this value of u_* in equation 7.17.

The minimum number of heights needed to determine $\partial T/\partial u$ is two.

If the heights are distinguished by subscripts 1 and 2, and u_* estimated from the neutral wind profile equation, equation 7.17 becomes

$$\mathbf{C} = \frac{k^2 \rho c_p (T_2 - T_1)(u_1 - u_2)}{\ln [(z_2 - d)/(z_1 - d)]^2} \qquad 7.20$$

An equation of this form was first derived to calculate water vapour transfer by Thornthwaite and Holzman and has been used in many subsequent studies of transfer in the turbulent boundary layer. Its main defect is the dependence on wind and temperature (or vapour pressure) at two heights only so that the estimate of flux is sensitive to the error in a single instrument or to local irregularities of the site. To minimize these sources of error, it is desirable to measure u and T at six or more heights above the surface so that good average values of $\partial T/\partial u$ and u_* can be derived from the profile.

8

Heat Transfer—(ii) Conduction

The more the earth is drained of heat, the colder becomes the moisture that is concealed in the ground.

Conduction is a form of heat transfer produced by a sharing of momentum between colliding molecules in a fluid, by the movement of free electrons in a metal, or by the action of inter-molecular forces in an insulator. In micrometeorology, conduction is important for heat transfer in the soil and through the coats of animals, but not in the free atmosphere where the effects of molecular diffusion are trivial in relation to mixing by turbulence.

STEADY STATE EQUATIONS

If the temperature gradient in a solid or motionless fluid is $\partial T/\partial z$, the rate of conduction of heat *per unit area* is proportional to the gradient and the constant of proportionality is called the **thermal conductivity** of the material k'. In symbols

$$\mathbf{G} = -k'\,(\partial T/\partial z) \qquad 8.1$$

where the negative sign is a reminder that heat moves in the direction of decreasing temperature. For a steady flow of heat between two parallel surfaces at T_2 and T_1 separated by a uniform slab of material with thickness t, integration of equation 8.1 gives

$$\mathbf{G} = -k'(T_2 - T_1)/t$$

More complex equations are needed to relate the heat transfer through cylinders and spheres to the temperature gradient. For a hollow cylinder with interior and exterior radii r_1 and r_2 at temperatures T_1 and T_2 respectively, the heat flux per unit area of the outer surface is

$$\mathbf{G} = \frac{k'(T_1 - T_2)}{r_2 \ln (r_2/r_1)} \qquad 8.2$$

and for a hollow sphere the corresponding flux is

$$\mathbf{G} = \frac{k'(T_1 - T_2)}{r_2(r_2/r_1 - 1)} \qquad 8.3$$

Equations 8.2 and 8.3 have been used to calculate heat transfer through parts of animals and birds that are approximately cylindrical or spherical.[9]

The form of equation 8.2 has important implications for the efficiency of insulation surrounding a cylinder or any object that is approximately cylindrical such as the trunk of a sheep or a human finger. The equation predicts that for a fixed value of the temperature difference across the insulation, the heat flow per unit length of the cylinder, $2\pi r_2 \mathbf{G}$ will be inversely proportional to $\ln (r_2/r_1)$. In the steady state, the rate of conduction must be equal to the convective heat loss from the outer surface of the insulation. If the air is at a temperature T_3, the convective heat loss will be

$$\mathrm{Nu}\,(k/2r_2)(T_2 - T_3) = 2\pi r_2 \mathbf{G} = k'(T_1 - T_2)/\ln (r_2/r_1)$$

where k is the thermal conductivity of air and Nu is the Nusselt number. By rearranging terms it can be shown that the heat loss per unit length of cylinder is

$$2\pi r_2 \mathbf{G} = \frac{2\pi k(T_1 - T_3)}{\{(k/k') \ln (r_2/r_1)\} + \{2/\mathrm{Nu}\}} \qquad 8.4$$

The Nusselt number is proportional to $(r_2)^n$ where n is 0·5 for forced and 0·75 for free convection. When r_2 increases, the heat loss $2\pi r_2 \mathbf{G}$ will therefore increase or decrease depending on which of the terms in curly brackets dominates the denominator. To find the break-even point where $2\pi r_2 \mathbf{G}$ is independent of r_2, equation 8.4 can be differentiated with respect to r_2 to give

$$\mathrm{Nu} = 2nk'/k$$

When the Nusselt number is greater than this critical value, the heat loss from the insulation increases as r_2 increases; when it is smaller than the critical value, the heat loss decreases with increasing r_2.

If the conductivity of air is taken as fixed and $n = 0{\cdot}6$ is an approximate mean value for mixed convection, the critical value of Nu depends only on k, the conductivity of the insulating material. For animal coats and for clothing, $k \simeq k'$, so the critical Nu is of the order of unity. In nature, Nusselt numbers of this size may be relevant to a furry caterpillar under a cabbage leaf on a calm night or to an animal in a burrow but, for most organisms freely exposed to the atmosphere, Nu will exceed 10. In general, therefore, the thermal insulation of animals will increase with the thickness of their hair or clothing.

The conductivity of fatty tissue on the other hand is about 12 times the conductivity of still air so the critical Nusselt number for insulation by subcutaneous fat is therefore about 14. When Nu is larger than 14, fat provides insulation in the conventional sense but in an environment where Nu is less than this critical value, a naked ape suffering from middle aged spread might have increasing difficulty in keeping warm as his girth increased.

INSULATION OF ANIMALS

All warm blooded animals are provided with insulation in the form of skin and subcutaneous tissue and hairy animals have an additional insulating layer of air between the skin and the surface at which radiation and convective heat exchanges take place.

Before measurements of insulation are reviewed, a link is needed between the unnecessarily clumsy and sometimes inaccurate units used by animal and human physiologists and the more elegant units of resistance adopted by many plant physiologists, ecologists and micrometeorologists. In the c.g.s. system for example, thermal conductance is often quoted, correctly but ambiguously, in units of kcal/m²/24 h/°C (i.e. kcal m^{-2} 24 h^{-1} $°C^{-1}$) but the inverse quantity, insulation (I) or specific resistance, has been quoted incorrectly and even more ambiguously in units of °C/Mcal/m²/24 h or even °C/cal m² 24 h instead of °C m² 24 h/Mcal.

As the insulation is a temperature difference per unit heat flux and per unit area, it is equivalent to the term $r/\rho c$ where ρc is a volumetric specific heat. To convert from units of insulation (e.g. °C m² 24 h $Mcal^{-1}$) to units of resistance (e.g. s cm^{-1}) and for comparison with the resistance of the boundary layer r_H, it is necessary to choose an arbitrary value of ρc_p for air, e.g. $2{\cdot}85 \times 10^{-4}$ cal cm^{-3} $°C^{-1}$ which is the value at 20°C. On this basis, a resistance $r = 1$ s cm^{-1} is equal to an insulation $r/\rho c_p =$ 4·00 °C m² 24 h $Mcal^{-1}$ and 1°C m² h $kcal^{-1}$ is very nearly 10 s cm^{-1}.

Another unit of insulation found mainly in human studies is the **clo** equal to 7·5°C m² 24 h $Mcal^{-1}$ and therefore equivalent to 1·86 s cm^{-1} or 0·39 cm of still air (putting $r_H = t/\kappa$, see p. 14). The clo was originally conceived as the insulation maintaining 'a resting man whose metabolism is 50 kcal m^{-2} h^{-1} indefinitely comfortable in an environment of 21°C, relative humidity less than 50%, and air movement 20 ft m^{-1}.' Specialized units of this type have few merits and tend to separate a subject from other related branches of science.

The insulation of animals has three components: a layer of tissue, fat and skin across which temperature drops from deep body temperature to mean skin temperature; a layer of relatively still air trapped within a coat of fur, fleece, feathers or clothing; and an outer boundary layer whose resistance is given by $d/(\kappa \, \mathrm{Nu})$ (p. 101). A comprehensive analysis of heat exchange in mammals would need to consider separately the amount of heat lost from the trunk, legs, head, etc. Because these appendages are usually less well insulated and are smaller than the trunk, they are capable of losing more heat *per unit area*. In practice, the loss of heat from appendages is usually small compared with the total loss from the rest of the body (although some animals subject to heat stress are believed to dissipate large amounts of heat through their ears or tails). Average values of insulation for different species can therefore be determined from measurements of metabolic heat production, external heat load, and the relevant mean temperature gradients.

Tissue

The insulation of tissue is strongly affected by the circulation of blood beneath the skin and the construction and dilation of blood vessels can

Table 8.1 Thermal resistances of animals
Peripheral tissue and coats

Tissue[12]	s cm^{-1}	
	Vaso-constricted	Dilated
Steer	1·7	0·5
Man	1·2	0·3
Calf	1·1	0·5
Pig (3 months)	1·0	0·6
Down sheep	0·9	0·3
Coats[12, 47]	s cm^{-1} per cm depth	per cent of still air
Air	4·7	100
Red fox	3·3	70
Lynx	3·1	65
Skunk	3·0	64
Husky dog	2·9	62
Merino sheep	2·8	60
Down sheep	1·9	40
Blackfaced sheep	1·5	32
Cheviot sheep	1·5	32
Ayrshire cattle: flat coat	1·2	26
erect coat	0·8	
Galloway cattle	0·9	19

change skin resistance by a factor of 2 to 3. For comparison with the thermal resistances of tissue, hair, and air, values of tissue insulation found in the literature have been multiplied by the volumetric heat capacity of air at 20°C. Table 8.1 shows values ranging from a minimum of 0·3 s cm^{-1} for dilated tissue to between 1 and 2 s cm^{-1} for vaso-constricted tissue.

Taken as an average over the whole body, the thermal conductivity of human skin during vaso-constriction appears to be about an order of magnitude greater than the conductivity of still air, i.e. about 0·2 to 0·3 W m^{-1} $°C^{-1}$, and measurements on fingers showed that the conductivity increased linearly with the rate of blood flow. The effective mean thickness of the skin during vaso-constriction is equivalent to about 2·5 cm of tissue or 0·25 cm of still air. These figures must conceal large local differences of insulation of the limbs and appendages depending on the thickness and nature of subcutaneous tissue and the degree of curvature.

Fur and feathers

The thermal resistance of animal coats has been measured by a number of workers and reported in units such as °C m^2 24 h $Mcal^{-1}$ or clo/inch. These units obscure an important physical fact: the thermal conductivity of hair, fleece and clothing is the same order of magnitude as the conductivity of still air. (A distinction made by others between 'still' air and 'dead air space' is based mainly on an arithmetical error. The two are physically identical by definition although in practice it is difficult to achieve a temperature gradient across a layer of still air without setting up a circulation of air by convection which increases in thermal conductivity.)

The thermal conductivity of air is $2{\cdot}5 \times 10^{-2}$ W m^{-1} $°C^{-1}$ at 20°C; or 4·8 s cm^{-1} for a layer of 1 cm; or 2·58 clo/cm; or 6·6 clo/in. Scholander and his colleagues[122] showed that the insulation per unit thickness of coat was remarkably uniform for a wide range of wild animals from shrews to bears (Fig. 8.1). The average insulation derived from his measurements is often quoted as 4 clo/in. meaning that 1 inch of fur had the same insulation as 4/6·6 or 0·6 inches of still air. In terms of still air, the efficiency of insulation is 60%. This implies that the air trapped between the hairs was not perfectly still, possibly because in this series of experiments a significant amount of heat was transferred by radiation and by free convection as well as by conduction. The skin surface was horizontal at 37°C and the air temperature was 0°C. For fur 5 cm thick, the Grashof number for free convection in the hair layer is 5×10^5. For flat plates the corresponding Nusselt number is about 2, implying that heat

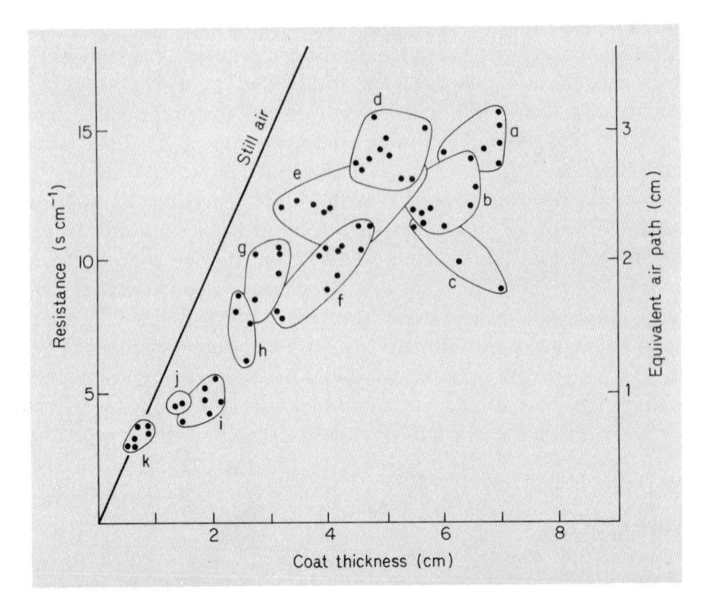

Fig. 8.1 Thermal resistance of animal coats as a function of their thickness (after Scholander *et al.*[122]).

a, dall sheep; b, wolf, grizzly bear; c, polar bear; d, white fox; e, reindeer, caribou; f, fox, dog, beaver; g, rabbit; h, marten; i, lemming; j, squirrel; k, shrew.

transfer by conduction and convection would proceed at twice the rate for conduction alone. Although the circulation of air in the fleece must be restricted by the presence of hairs, this calculation suggests that free convection may increase rates of heat transfer significantly, when a large temperature gradient exists across a layer of fur. The same argument can be applied to measurements by Hammel[47] who used horizontal samples of fur with a smaller temperature difference.

The existence of free convection in fur may explain why the erection of fur in response to cold has relatively little effect on thermal insulation. For example a flat calf coat normally 12 mm thick had an insulation[12] of 1·4 s cm^{-1}. After piloerection, the thickness of the hair almost doubled but its total resistance increased only to 1·9 s cm^{-1}. The resistance per centimetre decreased by 30%, presumably because free convection was more active in the deeper, thinner coat. Radiative exchange may also be an effective mode of heat transfer through relatively thin coats in which there are less than 10^3 hairs per cm^2. The efficiencies and corresponding resistances for a number of wild and domestic species are listed in Table 8.1. When the thickness of the hair layer is taken into account, the ratio of thermal resistance in the coat to the resistance in the skin is found to vary widely between species. In sheep and other animals with a deep layer of hair, the hair makes a relatively large contribution to the total resistance between the core and the external surface whereas the hair covering a man or a baby pig has little thermal significance.

A few attempts have been made to study the effect of windspeed and direction on the insulation of hair layers. The main effect of air movement is to penetrate the coat and destroy part of its insulation by decreasing the thickness of the still air trapped between hairs (Fig. 8.2). The relation

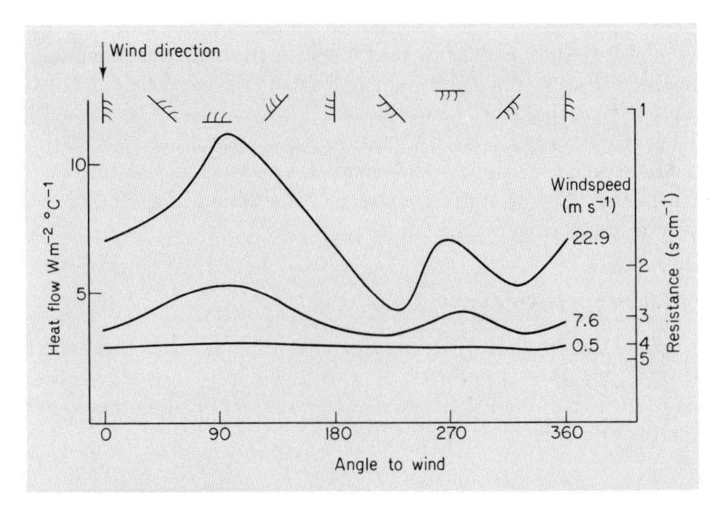

Fig. 8.2 Heat loss through and thermal resistance of the fur of newborn caribou from four samples from the backs of different animals. The upper part of the diagram shows the relation between wind direction and the natural set of the hairs (from Lentz and Hart[66]).

between insulation and wind speed is complicated by changes of coat thickness. Wind blowing in the same direction as the natural set of hair tends to decrease the thickness of the coat but wind in the opposite direction produces an irregular pattern of fluffing in some types of soft coat (Fig. 8.2). The effects of wind on coat insulation are therefore complex and difficult to quantify. Measurements on sheep[12] can be transformed to give the resistance of a sheep's fleece 1 cm deep as

$$r_f = 1{\cdot}5 - 0{\cdot}3V^{1/2} \text{ s cm}^{-1} \qquad 8.5$$

where V is the wind speed in m s^{-1}. A windspeed of 6 m s^{-1} is therefore enough to destroy half the insulation of the fleece. For a rabbit pelt about 2 cm deep, the resistance was

$$r_f = 3{\cdot}2 - 0{\cdot}47\, V^{1/2}\ \text{s cm}^{-1} \qquad 8.6$$

These average values for a whole animal conceal an important difference of insulation between windward and leeward surfaces. In one set of measurements[35] at an average windspeed of 5 m s^{-1}, the leeward insulation of a sheep fleece was three times the windward insulation but the average was consistent with the value predicted from equation 8.5.

HEAT CONDUCTION IN SOIL

Use of steady state equations for the conduction of heat in animal coats was justified by the small thermal capacity of coats and by the constancy of deep body temperature in mammals. The analysis of heat conduction in soils is more complex, partly because steady states are rare when a soil surface is exposed to annual and seasonal cycles of radiation and partly because changes in the water content or compaction of a soil may change its thermal properties profoundly.

Thermal properties of soil

By use of the symbol ρ for density and c for specific heat, the solid, liquid, and gaseous components of soil can be distinguished by subscripts s, l and g. If the volume fraction x of each component is expressed per unit volume of bulk soil

$$x_s + x_l + x_g = 1 \qquad 8.7$$

For a completely dry soil ($x_l = 0$), x_g is the space occupied by pores. In many sandy and clay soils it is between 0·3 and 0·4 and it increases with organic matter content, reaching 0·8 in peaty soils.

The bulk density of a soil ρ' is found by adding the weight of each component, i.e.

$$\rho' = \rho_s x_s + \rho_l x_l + \rho_g x_g = \sum(\rho x) \qquad 8.8$$

Because ρ_g for the soil atmosphere is much smaller than ρ_s or ρ_l, the term $\rho_g x_g$ can be neglected. When ρ_s and ρ_l are constant, soil density increases linearly with the liquid fraction x_l, but, in a soil which swells when it is wetted, the relation is not strictly linear.

The volumetric specific heat (J m^{-3}) is the product of bulk density ρ' and bulk specific heat c'. It can be found by adding the heat capacity of soil components to give

$$\rho' c' = \rho_s c_s x_s + \rho_l c_l x_l + \rho_g c_g x_g = \sum(\rho c x) \qquad 8.9$$

and this quantity increases linearly with water content in a non-swelling soil. The specific heat is therefore

$$c' = \frac{\sum \rho c x}{\sum \rho x}$$

Thermal properties of soil constituents and of three representative soils are listed in Table 8.2. Quartz and clay minerals, which are the main

Table 8.2 Thermal properties of soils and their components (after Van Wijk and de Vries[150])

	Water content x_1	Density ρ tonne m^{-3} (10^6 g m^{-3})	Specific heat c J g^{-1} °C^{-1}	Thermal conductivity k' W m^{-1} °C^{-1}	Thermal diffusivity κ' 10^{-6} m^2 s^{-1}
(a) *Soil components*					
Quartz		2·66	0·80	8·80	4·18
Clay minerals		2·65	0·90	2·92	1·22
Organic matter		1·30	1·92	0·25	1·00
Water		1·00	4·18	0·57	0·14
Air (20°C)		$1{\cdot}20 \times 10^{-3}$	1·01	0·025	20·50
(b) *Soils*					
Sandy soil	0·0	1·60	0·80	0·30	0·24
(40% pore space)	0·2	1·80	1·18	1·80	0·85
	0·4	2·00	1·48	2·20	0·74
Clay soil	0·0	1·60	0·89	0·25	0·18
(40% pore space)	0·2	1·80	1·25	1·18	0·53
	0·4	2·00	1·55	1·58	0·51
Peat soil	0·0	0·30	1·92	0·06	0·10
(80% pore space)	0·4	0·70	3·30	0·29	0·13
	0·8	1·10	3·65	0·50	0·12

solid components of sandy and clay soils, have similar densities and specific heats. Organic matter has about half the density of quartz but about twice the specific heat. As a result, most soils have volumetric specific heats between 2·0 and 2·5 J cm^{-3}. As the specific heat of water is 4·18 J cm^{-3}, the heat capacity of a dry soil increases substantially when it is saturated with water.

The dependence of thermal conductivity on water content is more complex. The thermal conductivity of a very dry soil may increase by an order of magnitude when a small amount of water is added because relatively large amounts of heat are transferred by the evaporation and condensation of water in the pores, a process described in detail by Rose. For a clay soil, for example, k' may increase from 0·3 to 1·8 W m^{-1} °C^{-1} when x_l increases from zero to 0·2. With a further increase of x_l from 0·2 to 0·4 the corresponding increase of k' is much smaller because the diffusion of vapour becomes increasingly restricted as more and more pores are filled with water. The conductivity of very wet soils is therefore almost independent of water content.

When water is added to a very dry soil, k' increases more rapidly than $\rho'c'$ initially so that the diffusivity $\kappa' = k'/\rho c$ also increases with water content. In a wet soil, however, the increase of k' with water content is much less rapid than the increase of $\rho'c'$ so that κ' decreases with water content. Between these two regimes, κ' reaches its maximum at a point where an increase of water content is responsible for equal fractional increases of k' and $\rho'c'$. Table 8.2 shows that sandy soils tend to have larger thermal diffusivities than other soil types because quartz has a much larger conductivity than clay minerals. Peat soils have the smallest diffusivities because the conductivity of organic matter is relatively small.

Formal analysis of heat flow

At a depth z below the soil surface, the downward flux of heat can be written

$$\mathbf{G}(z) = -k'(z)(\partial T/\partial z) \tag{8.10}$$

In any thin layer of thickness Δz, say, the difference between the flux entering the layer at level z and leaving at $z+\Delta z$ is $\mathbf{G}(z) - \mathbf{G}(z+\Delta z)$ or, in the notation of calculus, $\Delta z(\partial \mathbf{G}(z)/\partial z)$. The sign of this quantity determines whether there is net gain of flux or 'convergence' in the layer producing a local increase of soil temperature, or a net loss of flux or 'divergence' producing a fall in temperature. In general, the rate at which the heat content of the layer changes can be written $\partial(\rho'c'T\Delta z)/\partial t$ and this quantity must be equal to the change of flux with depth, i.e.

$$\frac{\partial \mathbf{G}(z)}{\partial z}\,\Delta z = \frac{\partial}{\partial z}\left(-k'\frac{\partial T}{\partial z}\right)\Delta z = -\frac{\partial(\rho'c'T)}{\partial t}\,\Delta z \tag{8.11}$$

For the special case in which the physical properties of the soil are constant with depth, the equation of heat conduction reduces to

$$\frac{\partial T}{\partial t} = \kappa'\frac{\partial^2 T}{\partial z^2} \tag{8.12}$$

Figure 8.3 is a graphical demonstration of this equation in terms of an imaginary temperature profile in a soil with constant diffusivity. Figure 8.4 shows the real change of temperature beneath a bare soil surface and beneath a crop.[149] Observed changes of temperature at different depths can be compared with the changes predicted from temperature gradients.

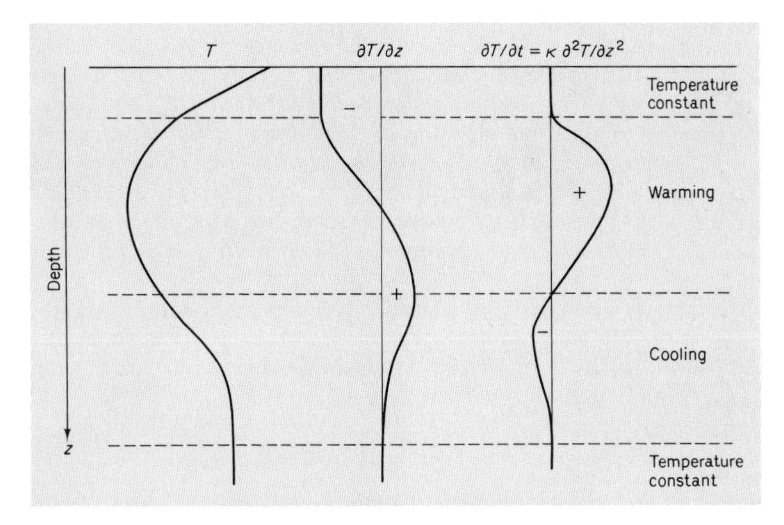

Fig. 8.3 Imaginary temperature gradient in soil (left-hand curve), and the corresponding first and second differentials of temperature with respect to depth, i.e. $\partial T/\partial z$ and $\partial^2 T/\partial z^2$. The second differential is proportional to the rate of temperature change $\partial T/\partial t$.

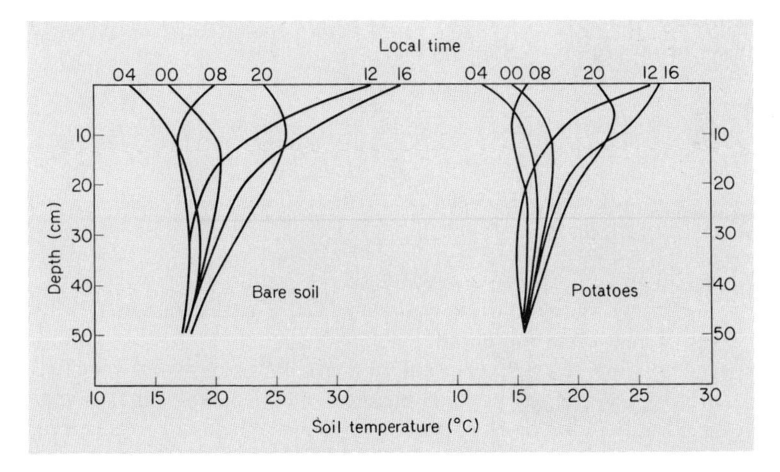

Fig. 8.4 Diurnal change of soil temperature measured below a bare soil surface and below potatoes (from van Eimern[149]).

In most soils, composition, water content and compaction change with depth, and in cultivated soils substantial changes often occur near the surface. Precise measurements of the bulk density and thermal conductivity of soils are therefore difficult to acquire *in situ*. The theory of heat transfer in soils has nevertheless been used to determine average thermal properties from an observed temperature regime as well as for predicting daily and seasonal changes of soil temperature. In most analyses, κ is assumed independent of depth but McCulloch and Penman[72] derived a solution of equation 8.12 when κ was a linear function of depth.

It is instructive to derive the temperature regime in a uniform soil by assuming the temperature of the soil-air interface oscillates sinusoidally during a daily or an annual cycle. If the temperature at time t and depth z is $T(z, t)$ the boundary condition becomes

$$T(0, t) = \bar{T} + A(0) \sin \omega t \tag{8.13}$$

where $\bar{T}$ is the mean surface temperature, and $A(0)$ is the amplitude of the oscillation at the surface so that the maximum and minimum temperatures are $\bar{T}+A$ and $\bar{T}-A$. The angular frequency of the oscillation ω is equal to 2π divided by the period of oscillation, i.e. for daily cycles $\omega=(2\pi/24)\ \mathrm{h}^{-1}$ with t in hours and for annual cycles $\omega=(2\pi/365)\ \mathrm{day}^{-1}$ with t in days.

The solution of equation 8.12 satisfying the boundary condition is

$$T(z, t) = T + A(z) \sin (\omega t - z/D) \tag{8.14}$$

where the amplitude at depth z is

$$A(z) = A(0)\, \mathrm{e}^{-z/D} \tag{8.15}$$

and

$$D = (2\kappa/\omega)^{1/2} \tag{8.16}$$

Several important features of conduction in soils can be related to the value of D as follows:

(i) at a depth $z=D$, the amplitude of the temperature wave is e^{-1} or 0·37 times the amplitude at the surface. For this reason D is often called the 'damping depth';

(ii) the position of any fixed point on a temperature wave is specified by a fixed value of the phase angle $(\omega t - z/D)$, e.g. the angle is $\pi/2$ for maximum temperature and $-\pi/2$ for minimum temperatures. Differentiation of the simple equation $\omega T - z/D = \text{constant}$ gives $\partial T/\partial z = \omega D$ and this is the velocity with which temperature maxima and minima appear to move downwards into the soil;

(iii) at a depth $z = \pi D$, the phase angle is π less than the angle at the surface, i.e. the temperature wave is exactly out of phase with the wave at the surface. When the surface temperature reaches a maximum, the temperature at πD reaches a minimum and vice versa;

(iv) by differentiating equation 8.12 with respect to z and putting $z = 0$, it can be shown that the heat flux at the surface at time t is

$$\mathbf{G}(0, t) = \frac{\sqrt{2}\, A(0)k \sin(\omega t + \pi/4)}{D}$$

The maximum heat flux is $\sqrt{2}\, A(0)k/D$ which is the flow of heat that would be maintained through a slab of soil with thickness $\sqrt{2}\, D$ if one face were maintained at the maximum and the other at the minimum temperature of the surface. The quantity $\sqrt{2}\, D$ can therefore be regarded as an effective depth for heat flow. (Note that the flux reaches a maximum $\pi/4$ or one eighth of a cycle before the temperature, i.e. 3 hours for the diurnal wave and $1\frac{1}{2}$ months for the annual wave);

(v) the amount of heat flowing *into* the soil during one half cycle is found by integrating $\mathbf{G}(0)$ from $\omega T = -\pi/4$ to $+3\pi/4$ and is $\sqrt{2}\, \rho c D A(0)$. This is the amount of heat needed to raise through $A(0)$°C a layer of soil equal to the effective depth $\sqrt{2}\, D$.

Values of D for three types of soil are plotted in Fig. 8.5 for daily and annual cycles and as a function of volumetric water content. In sandy and clay soils, D increases rapidly when x_l increases from 0·0 to 0·1 reaching values between 12 and 18 cm for the daily cycle. For the peat soil, D lies between 3 and 5 cm over the whole range of water content, consistent with the slow heating or cooling of organic soil in response to changes of radiation or of air temperature. Corresponding values for the out-of-phase depth πD, the effective depth and the rate of penetration of the temperature wave can be read from the appropriate axes.

Figure 8.6 shows the type of record from which a damping depth and heat fluxes can be estimated. From the amplitude of the temperature waves at 25 and 30 cm, it can be shown that the surface amplitude $2A(0)$ was about 20°C. The value of D is 10 cm giving $\kappa = 3{\cdot}6 \times 10^{-3}$ cm^2 s^{-1} from equation 8.16 and as the soil was a sandy loam, these values imply that it was very dry (see Fig. 8.3 and Table 8.2). The volumetric heat capacity for a dry sandy soil is about 1·6 MJ m^{-3} °C^{-1} and the thermal conductivity $k = \kappa \rho c$ would therefore be about 0·6 W m^{-1} °C^{-1}. If the effective depth is taken as 14 cm ($\sqrt{2}\, D$), the maximum heat flux into the soil is $40k/\sqrt{2}\, D = 170$ W m^{-2} and the amount of heat stored in the soil during a half cycle is $\sqrt{2}\, D \rho c A(0) = 4{\cdot}6$ MJ m^{-2}.

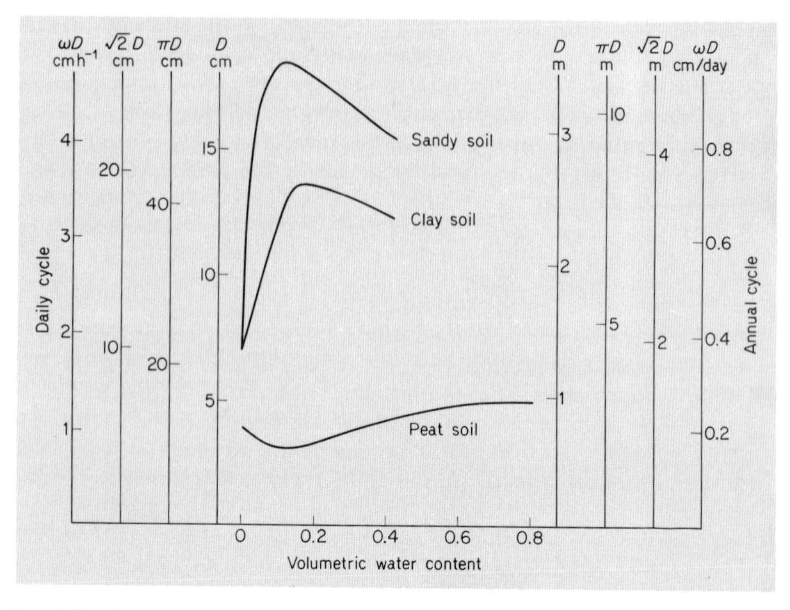

Fig. 8.5 Change of damping depth and related quantities for three soils over a wide range of water contents. Left-hand axes refer to a daily and right-hand axes to an annual cycle (data from van Wijk and de Vries[150]).

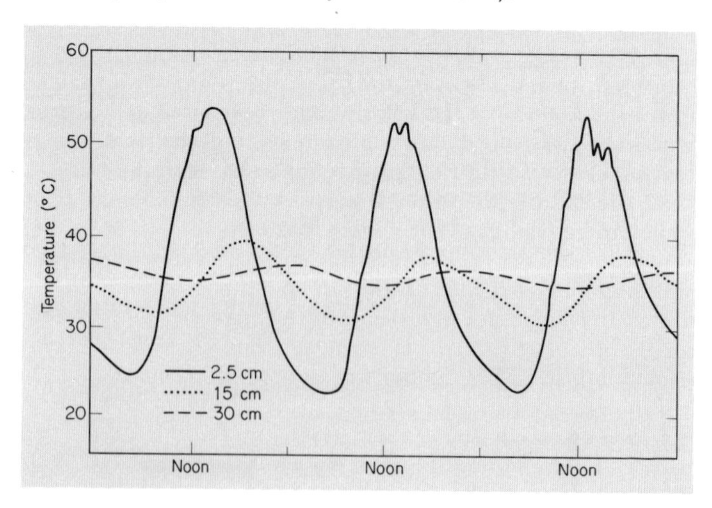

Fig. 8.6 Diurnal course of temperature at three depths in a sandy loam beneath a bare uncultivated surface; Griffith, New South Wales, 17–19 January, 1939 (from Deacon[30] after West).

In this analysis, the surface amplitude $A(0)$ has been treated as the independent variable in the system. In practice $A(0)$ depends on the heat balance of the soil surface and on the relative thermal properties of soil and atmosphere. To compare the behaviour of different soils exposed to the same weather it is necessary to begin by calculating the amplitude of the heat flux $\mathbf{G}(0)$ (see p. 160). The surface amplitude and the soil temperature distribution can then be derived as a function of D. In practice, it is usually much easier to measure temperature and soil heat flux directly than to attempt calculations based on the idealized model of a uniform soil.

Modification of thermal regimes

Many biological processes depend on soil temperature: the metabolism and behaviour of microorganisms and many invertebrates; the germination of seeds and extension of root systems; tillering and shoot extension. Because it is difficult to observe the behaviour of an undisturbed root system or of the fauna and flora in a natural soil, relatively little is known about physiological responses to changes of soil temperature or about the behavioural significance of soil temperature gradients. Ignorance of fundamental processes has not prevented agronomists, horticulturists and foresters from developing empirical methods of modifying the thermal regime of soils to help the establishment and growth of crops and trees. Their methods include mulching soils with layers of organic matter in the form of peat or straw to reduce heat losses in winter, covering peat soils with a layer of sand to inhibit evaporation and to reduce the risk of frost at the soil-air interface; tillage to increase soil surface temperature in spring; and the use of black or white powders to raise or lower the temperature of the surface by changing its reflectivity.

Apart from direct intervention by man, the temperature regime of any soil is profoundly modified by the growth of vegetation because the surface becomes increasingly shaded as the canopy develops. The presence of shade reduces the maximum and increases the minimum temperature recorded at any depth and there is usually a small decrease in average temperature. Although this effect is well documented, the implications for root and rhizosphere activity have not been explored.

9

Mass Transfer

There is no need of words to show that the sea, rivers, and springs are constantly being replenished with a perennial flow of water.

Two modes of diffusion are responsible for the exchange of matter between organisms and the air surrounding them. *Molecular* diffusion operates within organisms (e.g. in the lungs of an animal or in the substomatal cavities of a leaf) and in a thin skin of air forming the boundary layer that surrounds the whole organism. In the free atmosphere, transfer processes are dominated by the effects of *turbulent* diffusion although molecular diffusion continues to operate and is responsible for the ultimate degradation of turbulent energy into heat.

Turbulence is ubiquitous in the atmosphere except very close to the earth's surface on very calm nights and the turbulent transfer of water vapour and carbon dioxide is of paramount importance for all higher forms of life. As a measure of effectiveness, the amount of carbon dioxide absorbed by a healthy green crop in one day is equivalent to *all* the CO_2 between the canopy and a height of 30 m. In practice, although the concentration of carbon dioxide in the atmosphere decreases between sunrise and sunset as a result of photosynthesis, this depletion rarely exceeds 15% of the mean concentration at the surface. These figures imply that turbulent transfer enables vegetation to extract CO_2 from the lowest 200 m of the atmosphere and probably from much greater heights. A small diurnal change has been reported at 500 m.

The process of mass transfer will now be described in terms of diffusion across boundary layers, through porous septa and within the free atmosphere.

NON-DIMENSIONAL GROUPS

Mass transfer to or from objects suspended in a moving airstream is analogous to heat transfer by convection and is conveniently related to a non-dimensional parameter similar to the Nusselt number of heat transfer theory. This is the **Sherwood number** defined by the equation

$$\mathbf{F} = \mathrm{Sh}\, D(\chi_s - \chi)/d \qquad 9.1$$

where $\mathbf{F}$ = mass flux of a gas per unit surface area (e.g. $\mathrm{g\ m^{-2}\ s^{-1}}$);
χ_s, χ = mean concentration of gas at the surface and in the free atmosphere (e.g. $\mathrm{g\ m^{-3}}$);
D = molecular diffusivity of the gas in air (e.g. $\mathrm{m^2\ s^{-1}}$).

As

$$\mathrm{Sh} = \frac{\mathbf{F}}{D(\chi_s - \chi)/d}$$

the Sherwood number can be defined as the ratio of actual mass transfer $\mathbf{F}$ to the rate of transfer that would occur if the same concentration difference were established across a layer of still air of thickness d. The corresponding resistance to mass transfer is derived by comparing equation 9.1 with

$$\mathbf{F} = (\chi_s - \chi)/r$$

giving $r = d/(D\ \mathrm{Sh})$ (cf. $r_H = d/(K\ \mathrm{Nu})$). Resistances and diffusion coefficients for water vapour and carbon dioxide will be distinguished by subscripts and are related by $r_V = d/(D_V\ \mathrm{Sh})$ and $r_C = d/(D_C\ \mathrm{Sh})$.

Just as the Nusselt number for forced convection is a function of Vd/ν (Reynolds number) and ν/κ (Prandtl number), the Sherwood number is the same function of Vd/ν and the ratio ν/D which is known as the **Schmidt number** and is abbreviated to Sc. For example, the Sherwood number for mass exchange at the surface of a flat plate is

$$\mathrm{Sh} = 0{\cdot}66\ \mathrm{Re}^{1/2}\ \mathrm{Sc}^{1/3} \qquad 9.2$$

cf.

$$\mathrm{Nu} = 0{\cdot}66\ \mathrm{Re}^{1/2}\ \mathrm{Pr}^{1/3} \qquad 9.3$$

The presence of the term $0{\cdot}66\ \mathrm{Re}^{1/2}$ in both expressions is a consequence of a fundamental similarity between the molecular diffusion of heat, mass and momentum in laminar boundary layers and the numbers $\mathrm{Sc}^{1/3}$ and $\mathrm{Pr}^{1/3}$ take account of differences in the effective thickness of the boundary layers for mass and heat.

For any system in which heat transfer is dominated by forced convection, the relation between Sh and Nu is given by dividing equation 9.2 by 9.3:

$$\mathrm{Sh} = \mathrm{Nu}\ (\mathrm{Sc}/\mathrm{Pr})^{1/3} = \mathrm{Nu}\ (\kappa/D)^{1/3} \qquad 9.4$$

The ratio κ/D is sometimes referred to as a **Lewis number** (Le). In air at 20°C, $(\kappa/D)^{1/3}$ is 0·96 for water vapour and 1·14 for CO_2 (see Table A.2, p. 220). The corresponding ratios of resistances are

$$r_V/r_H = (\kappa/D_V)^{2/3} = 0{\cdot}93 \qquad 9.5a$$

$$r_C/r_H = (\kappa/D_C)^{2/3} = 1{\cdot}32 \qquad 9.5b$$

In free convection, the circulation of air round a hot or cold object is determined by differences of air density produced either by temperature gradients or by vapour concentration gradients or by a combination of both. If the Nusselt number is related to the Grashof and Prandtl numbers by $\mathrm{Nu} = B\ \mathrm{Gr}^n\ \mathrm{Pr}^m$, the Sherwood number will be $\mathrm{Sh} = B\ \mathrm{Gr}^n\ \mathrm{Sc}^m = \mathrm{Nu}\ \mathrm{Le}^m$ where m is $\frac{1}{4}$ in the laminar regime and $\frac{1}{3}$ in the turbulent regime. To calculate the Grashof number, it is convenient to replace the difference between the surface and air temperature $T_0 - T$ by the difference of virtual temperature (p. 7). If e_0 and e are vapour pressures at the surface and in the air and p is air pressure, the gradient of virtual temperature is

$$T_{v0} - T_v = T_0(1+0{\cdot}61e_0/p) - T(1+0{\cdot}61e/p)$$
$$= (T_0 - T) + 0{\cdot}61(e_0T_0 - eT)/p$$

where temperatures are expressed in K. The importance of the vapour pressure term when T is close to T_0 can be illustrated for the case of a man covered with sweat at 33°C and surrounded by still air at 30°C and 20% relative humidity. Then $e_0 = 50{\cdot}3$ mbar and $e = 8{\cdot}5$ mbar. The term $T_0 - T$ is 3 K, $0{\cdot}61(e_0T_0 - eT)/p$ is 7·8 K and the difference of virtual temperature is 10·8 K. The size of the Grashof number allowing for the difference in vapour pressure is 3·6 times the number calculated from the temperature difference alone. The corresponding error in calculating a Nusselt or Sherwood number (proportional to $\mathrm{Gr}^{1/4}$) is about -38%.

The same type of calculation is used to determine atmospheric stability when temperature and water vapour concentration are both functions of height.

MEASUREMENTS OF MASS TRANSFER

Plane surfaces

For laminar flow over smooth flat plates, the Sherwood number for water vapour, $0{\cdot}57\ \mathrm{Re}^{0{\cdot}5}$, is shown by the continuous line in Fig. 9.1. Powell[108] obtained a very similar relation $\mathrm{Sh} = 0{\cdot}41\ \mathrm{Re}^{0{\cdot}56}$ for circular discs with diameters between 5 and 22 cm parallel to the wind (pecked line). Thom[137] measured evaporation from filter paper attached to the model bean leaf described on p. 83 and used bromobenzene and methyl salicylate as well as water to get a range of diffusion coefficients from 0·054 to 0·24 $cm^2 s^{-1}$. At windspeeds exceeding 1 $m s^{-1}$, the mass transfer of all three vapours was described by $Sh = 0{\cdot}7\ Re^{0.5}\ Sc^{0.33}$ within a few per cent of the predicted value. The measurements for water vapour plotted in Fig. 9.1 show that when the windspeed was less than 1 $m s^{-1}$

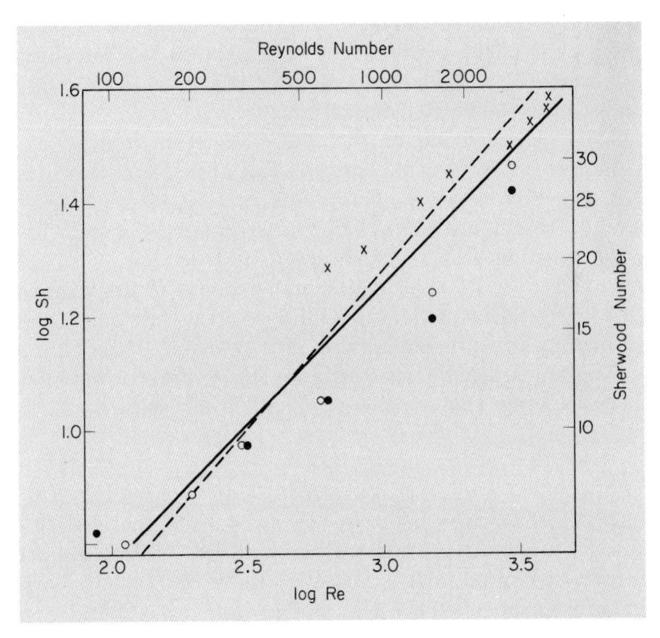

Fig. 9.1 Relation between Sherwood and Reynolds numbers for plates parallel to the airstream. Continuous line, standard relation $Sh = 0{\cdot}57\ Re^{0.5}$; pecked line, measurements on discs by Powell[108]; X, measurements on model bean leaf by Thom[137]; 0, measurements on replicas of alfalfa and Cocksfoot leaves by Impens.

($Re < 2800$), the Sherwood numbers were larger than the predicted value, possibly because the rate of mass transfer was increased by differences of density in the air surrounding the leaf.

Impens determined the rate of evaporation from blotting paper replicas of simulated leaves of alfalfa (*Medicago sativa*) and Cocksfoot (*Dactylis glomerata*) and his unpublished results are shown as Sherwood numbers in Fig. 9.1. The values of Sh are close to the values predicted for flat plates both at high and at low Reynolds numbers but are about 25% smaller than the standard relation at intermediate values of Re. Impens[54] also found that when blotting paper replicas of *Phaseolus* leaves were exposed at different heights in a stand of *Phaseolus*, transfer coefficients were close to the values predicted for flat plates of the same size.

To summarize the results of these experiments, there is no evidence to suggest that mass transfer coefficients for leaf replicas in a wind tunnel or in the field are substantially different from the values expected for flat plates of comparable size and shape. Sherwood numbers for real leaves may be 20 or 30% larger than for leaf replicas because transfer rates are increased by the effects of fluttering or roughness but the anomalously large heat transfer coefficients reported for maize leaves (p. 107) have no counterpart in mass transfer measurements.

Several workers have shown that the evaporation from a disc or leaf replica changes with the angle between the surface and the direction of the wind. For real leaves, however, leaf angle is mainly important for determining the amount of radiation intercepted by a leaf. When the amount of radiation absorbed by a leaf is independent of its angle, the transpiration rate will appear to be independent of angle because the boundary layer resistance is nearly always a relatively small part of the diffusion path. The diffusion resistance of stomata, discussed at the end of this chapter, is seldom less than 1 s cm^{-1} whereas boundary layer resistances are often about 0·2 to 0·4 s cm^{-1}.

Cylinders

For Reynolds numbers between 10^3 and 5×10^4, the Nusselt number for cylinders can be expressed as $Nu=0{\cdot}26\ Re^{0\cdot6}\ Pr^{0\cdot33}$ and the corresponding Sherwood number is $Sh=0{\cdot}26\ Re^{0\cdot6}\ Sc^{0\cdot33}$ or $Sh=0{\cdot}22\ Re^{0\cdot6}$. Powell's measurements on the evaporation from a wet cylinder in a wind tunnel fit this relation closely and it is shown by the continuous line in Fig. 9.2. For the more restricted range of Reynolds number from 4×10^3 to 4×10^4, the relation $Nu=0{\cdot}17\ Re^{0\cdot62}$ is often used. The corresponding Sherwood number $Sh=0{\cdot}16\ Re^{0\cdot62}$ is shown by the pecked line in Fig. 9.2.

Rapp[112] established close agreement between measurements of the evaporative loss from nude men covered with sweat and values predicted from the Sherwood number for a cylinder of appropriate diameter. The units of his calculations have been transformed to show this agreement in Fig. 9.2.

Spheres

The Nusselt number for a sphere can be expressed in the form $0{\cdot}34\ Re^{0\cdot6}$ and the corresponding Sherwood number is $0{\cdot}34\ Re^{0\cdot6}\ Le^{1/3}$ or

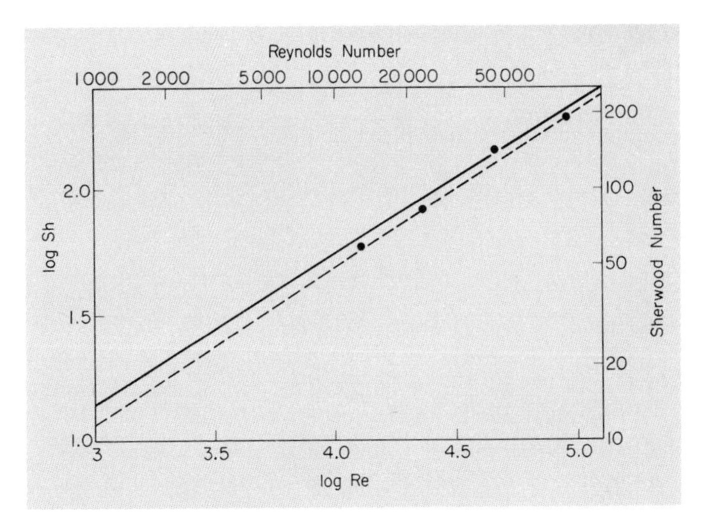

Fig. 9.2 Relations between Sherwood and Reynolds numbers for a wet cylinder at right-angles to the airstream. Continuous line, $Sh = 0{\cdot}22\ Re^{0{\cdot}6}$; pecked line, $Sh = 0{\cdot}16\ Re^{0{\cdot}62}$. The points were calculated by Rapp[112] from measurements by Kerslake on a man covered with sweat.

$0{\cdot}32\ Re^{0{\cdot}6}$ for water vapour. Analysis of Powell's measurements[108] of evaporation from wet spheres gives $Sh = 0{\cdot}26\ Re^{0{\cdot}59}$, about 20% less than the value predicted from heat transfer rates.

Mass transfer by ventilation

When mass transfer is induced by the ventilation of a system, an equation relating the mass flux to an appropriate potential gradient can be used to define a transfer resistance consistent with the values of diffusion resistance already discussed. Two relevant examples are the exchange of carbon dioxide between the air inside and outside a glasshouse and the loss of water vapour by evaporation from the lungs of an animal.

If the air in a glasshouse is well stirred so that the volume concentration of CO_2 in the internal air has a uniform value ϕ_i ($m^3\ CO_2\ m^{-3}$ air) when the external concentration is ϕ_e, the rate at which plants in the glasshouse absorb CO_2 from the external atmosphere can be written as

$$Q = \rho_c v N(\phi_e - \phi_i)\ g\ h^{-1} \tag{9.6}$$

where v is the volume of air in the house (m^3), N is the number of air changes per hour and ρ_c is the density of CO_2 ($g\ m^{-3}$). Dividing both sides of equation 9.6 by the floor area A gives the flux of CO_2 per unit floor area

$$\mathbf{F} = Q/A = \rho_c v N(\phi_e - \phi_i)/A$$

The resistance to CO_2 diffusion can be defined by writing

$$\mathbf{F} = \rho_c(\phi_e - \phi_i)/r_c$$

and comparison of the two equations gives

$$r_v = A/vN$$
$$= (N\bar{h})^{-1} \qquad 9.7$$

where $\bar{h}$ is the mean height of the house. For example, if N is 10 air changes per hour and $\bar{h} = 3$ m, r_v is 1/30 h m^{-1} or 1·2 s cm^{-1}, comparable in size with boundary layer and stomatal resistances.

Similarly if $\dot{V}$ is the minute volume of an animal's respiratory system and A is the area of skin surface, the loss of water per unit skin area is

$$\mathbf{F} = \dot{V}(\chi_s(T_b) - \chi_0)/60A$$

where $\chi_s(T_b)$ is the water vapour concentration of air saturated at deep body temperature and χ_0 is the concentration in the environment, both expressed in g m^{-3}. Then

$$r_v = 60A/\dot{V}$$

For a man at rest $\dot{V} = 10^{-2}$ m^3 min^{-1} and if $A = 1·7$ m^2, r_v is 10^4 s m^{-1} or 100 s cm^{-1}, i.e. about two orders of magnitude larger than common values of the boundary layer resistance for a sweating nude figure. Even during very rapid respiration when $\dot{V}$ may reach 10^{-1} m^3 min^{-1}, the diffusion resistance for respiration will exceed the boundary layer resistance by an order of magnitude and, when the skin is covered with sweat, the loss of water by evaporation from the lungs will be much smaller than the cutaneous evaporation rate.

Mass transfer by turbulence

The exchange of water vapour and carbon dioxide by atmospheric turbulence is a process analogous to the turbulent transfer of heat discussed on pages 116 to 118. In a boundary layer where the diffusion coefficients for mass and momentum are equal, the gradient of mass per unit value $\partial c/\partial z$ is related to the gradient of momentum per unit volume $\partial(\rho u)/\partial z$ by the identity

$$-\frac{\partial c/\partial z}{\mathbf{F}} = \frac{\partial(\rho u)/\partial z}{\tau}$$

where $\boldsymbol{\tau}$ is the flux of momentum and $\mathbf{F}$ is the mass flux in consistent units. Provided the vertical mass and momentum fluxes are constant with height, this relation implies that c is a linear function of u and by re-arranging terms and putting $\boldsymbol{\tau}=\rho u_*^2$ it can be shown that

$$\mathbf{F} = -(\partial c/\partial u)u_*^2$$

In practice, the gradient $\partial c/\partial u$ can be determined by measuring the concentration c of water vapour or carbon dioxide at a number of heights, say six, above a uniform surface and within the equilibrium boundary layer (p. 95) and plotting c against u to give a straight line whose slope is $\partial c/\partial u$. In neutral conditions, the friction velocity is determined from the wind gradient alone using equation 6.11 and in non-neutral conditions it is determined from the wind gradient and $\partial T/\partial u$ using equation 7.19. An example of this procedure is discussed in the last chapter.

MASS TRANSFER THROUGH PORES

When leaves transpire, water evaporates from cell walls and escapes to the atmosphere by diffusing into substomatal cavities, through stomatal pores, and finally through the leaf boundary layer into the free atmosphere. During photosynthesis, molecules of carbon dioxide follow the same diffusion path in the opposite direction. The resistance offered by the boundary layer to the diffusion of a gas depends on leaf dimensions and on windspeed whereas the resistance of stomatal pores depends only on the geometry, size and spacing of the pores and on associated anatomical features.

Meidner and Mansfield[77] tabulated stomatal populations and dimensions for 27 species including crop plants, deciduous trees and evergreens. The leaves of many species have between 100 and 200 stomata per mm^2 distributed on both the upper and lower epidermis (amphistomatous leaf) or on the lower surface only (hypostomatous leaf). The length of the pore is commonly between 10 and 30 μm and the area occupied by a complete stoma, including the guard cells responsible for opening and shutting the pore, ranges from 25×17 μm in *Medicago sativa* to 72×42 μm in *Phyllitis scolopendrium*.

Because stomata tend to be smaller in leaves where they are more numerous, the fraction of the leaf surface occupied by pores does not vary much between species and is about 1% on average for a pore width of 6 μm. There is much greater variation in the geometry of pores: the stomata of grasses are usually long, narrow, and aligned in rows parallel to the midrib whereas the elliptical stomata of sugar beet (*Beta vulgaris*) and broad bean (*Vicia faba*) are randomly oriented but uniformly dispersed over the epidermis.

The network of resistances in Fig. 9.3 is an electrical analogue for the diffusion of water vapour between the intercellular spaces and the external air. The calculation of boundary layer resistance r_V has already been discussed: values of 0·3 to 1 s cm^{-1} are expected for small leaves in a light wind. Many mesophytes have minimum stomatal resistances in the range 1 to 2 s cm^{-1} but values as small as 0·5 s cm^{-1} and as large as 4·8 s cm^{-1} have been reported for *Beta vulgaris* and *Phaseolus vulgaris*

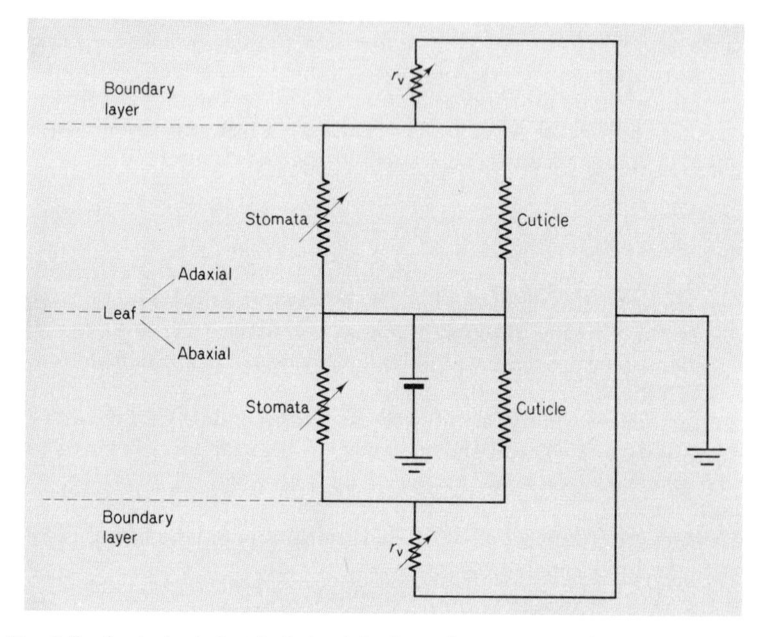

Fig. 9.3 Equivalent electrical circuit for loss of water vapour from a leaf by diffusion through the stomata and cuticle of the upper and lower epidermis.

respectively. Xerophytes have larger minimum resistances up to 30 s cm^{-1}. Cuticular resistances range from 20 to 60 s cm^{-1} in mesophytes and from 40 to 400 s cm^{-1} in xerophytes. In both types of plant, the resistance of the cuticle is usually so much larger than the stomatal resistance that its role in water vapour and CO_2 transfer can generally be ignored.

Resistance calculations

Electrical analogues provide a useful way of visualizing the process of diffusion from the intercellular spaces of a leaf through stomatal pores to the external boundary layer. Figure 9.4 shows a very simple analogue in which electrical current flows through a thin sheet of metal between two electrodes XX and YY represented by lines of dots. Suppose that the resistance of the sheet is 6 ohms so that the drop in voltage between the two electrodes is 6 volts when the current flowing between them is 1 amp. The pecked lines represent points of equal voltage across the sheet and

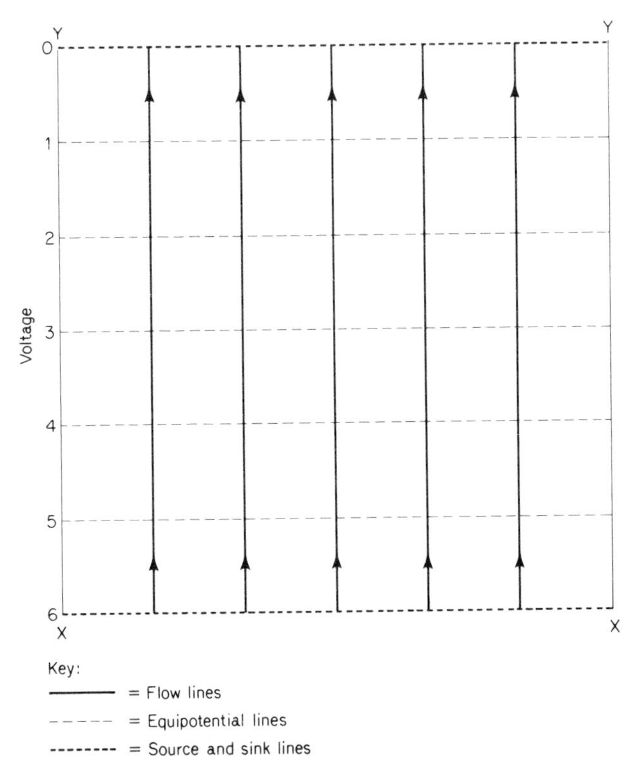

Fig. 9.4 Two-dimensional electrical analogue of diffusion between parallel planes XX and YY. The continuous lines represent flow of current and the pecked lines join points of equal potential.

from the symmetry of the system these equipotential lines must be parallel to the electrodes. Between each pair of adjacent lines the voltage drops by 1 volt so the material between them has a resistance of 1 ohm. The path of the current is represented by the bold lines at right angles to the equipotential lines and the arrows show the direction of current flow.

In this example, the equipotential lines form a very simple pattern which could be drawn from first principles. The distribution of lines in a more complicated system can be determined by using sheets of paper which have been impregnated with graphite so that they conduct electricity although the resistance per unit length is relatively large. Low resistance electrodes of appropriate shape are drawn on the paper with

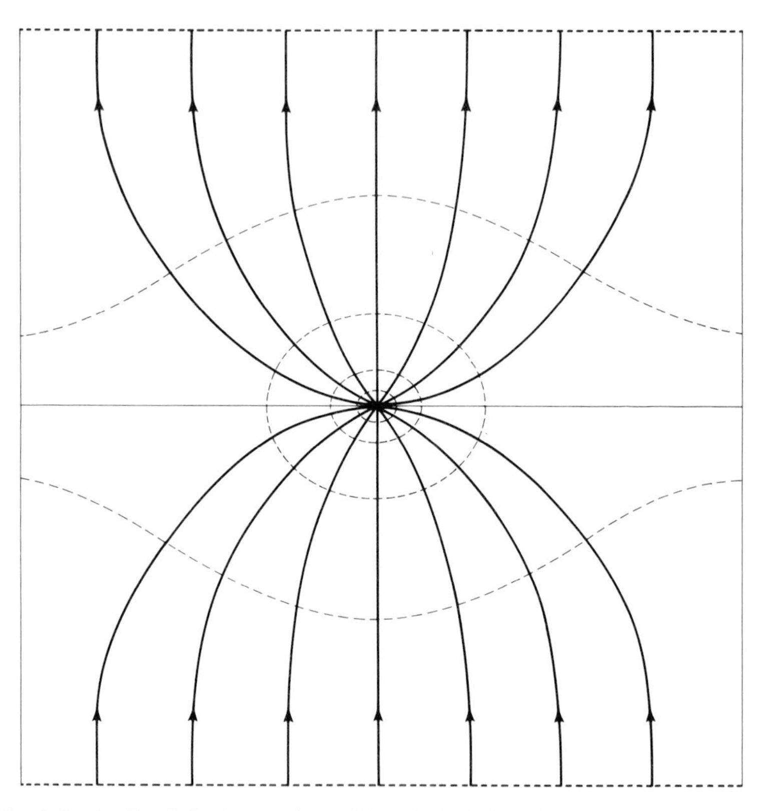

Fig. 9.5 As Fig. 9.4 when a plate with a single hole is introduced between the two planes. The plate is supposed to be infinitely thin.

silver paint and the shapes of the equipotential lines are determined with a field plotter. Figure 9.5 shows the pattern of lines on a sheet which had two parallel electrodes drawn 60 cm long and 60 cm apart. The sheet was cut along a line midway between the electrodes so that in the centre of the sheet the flow of current was confined to a neck of conducting paper only 7 mm wide. The way in which the flow of current converged on the neck is shown by the lines of flow drawn at right angles to the equipotential lines. The additional resistance imposed by cutting the paper can be determined by counting the equipotential lines which are separated by unit resistance as in Fig. 9.4. For the complete sheet there were 6 lines (counting one electrode as a line) whereas the cut sheet was found to have 16 lines (some were omitted for clarity). The cut therefore introduced 10 additional units of resistance or 5 units for each side of the neck.

Figure 9.5 can be regarded as a two-dimensional analogue for the diffusion of a gas through a circular hole with a diameter d in a plate whose thickness is much less than d. According to the theory of gaseous diffusion in three dimensions, the resistance of each side of a circular hole is $r_h = \pi d/8D$ where D is the diffusion coefficient of the gas. This quantity corresponds to the 5 units of electrical resistance in the two-dimensional analogue. The resistance of a laminar boundary layer of thickness t is $r_a = t/D$ and this corresponds to 3 units of resistance in the analogue.

The next stage in developing a realistic stomatal analogue is to introduce a pore of length l comparable with its diameter d. When an analogue was cut from conducting paper with $l = 10$ mm, $d = 7$ mm, the number of equipotential lines increased from 16 to 26. The additional 10 lines appeared within the pore at right angles to the walls and the rest of the analogue was identical to Fig. 9.5. In three dimensions, the diffusion resistance equivalent to the lines within a uniform pore is $r_p = l/D$. The total resistance of a pore r_t is found by adding r_p to the resistance r_h at either end of the pore, i.e. $r_t = r_p + 2r_h$. For stomata, r_h is often smaller than r_p and is therefore referred to as an 'end-correction'.

Figure 9.6 shows the cross section of a real pore and substomatal cavity for a leaf of *Zebrina pendula*. As the cross section is not uniform, the resistance of the pore cannot be calculated accurately without knowing the shape of the cross section and its area A at different distances from the end of the pore. An approximate value of r_p can be obtained from the length and mean diameter d of a pore with circular cross section or from the axes of an elliptical pore. The end-correction for a circular pore is usually assumed to be $2r_h = \pi d/4D$ where d is a representative diameter but Fig. 9.6 shows that it would be incorrect to apply a conventional end-correction to the inner end of the pore. If the substomatal activity is assumed to be lined with cell walls from which water is evaporating (represented by the dotted electrode), the resistance of the outer end of the pore is equivalent to 6 equipotential lines, but the inner end has only 2 lines. For many leaves, $r_t = r_p + r_h$ is probably a better estimate of the resistance of a single pore than $r_t = r_p + 2r_h$.

Finally, the resistance of a multi-pore system can be estimated. As mesophytes usually have about 100 stomata per mm^2 their average separation is about 0·1 mm or 100 μm, an order of magnitude larger than

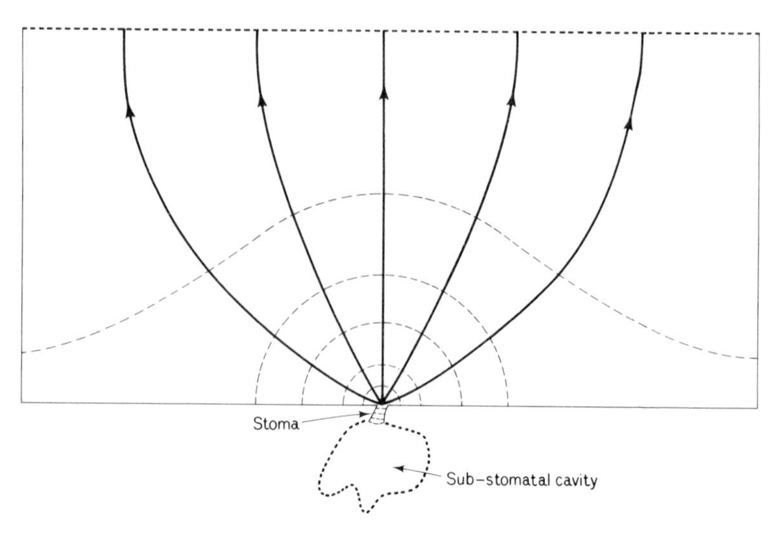

Fig. 9.6 Electrical analogue of diffusion of water vapour from single stomatal pore. Note the absence of equipotential lines in the sub-stomatal cavity showing that the 'end-correction' can be neglected at the inner end of the pore.

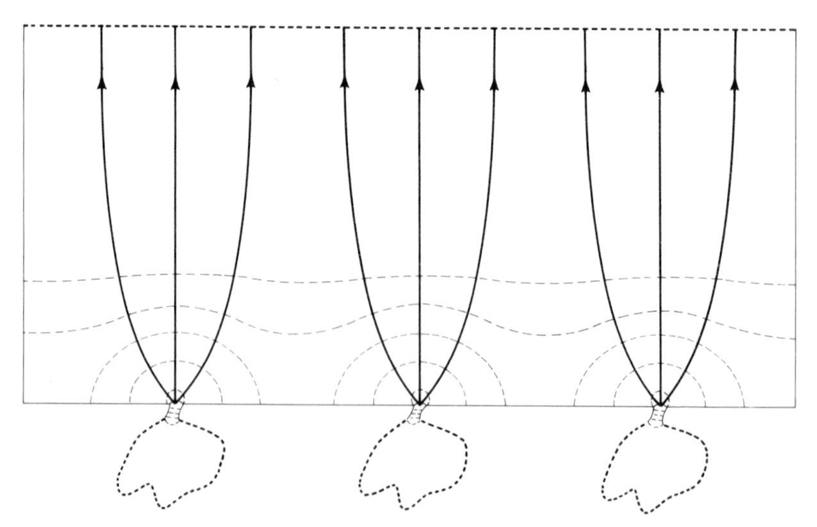

Fig. 9.7 Electrical analogue for a leaf epidermis showing three pores. Note the merging of the equipotential lines which decrease the effective end-correction for the outer end of each pore.

the maximum diameter of the pore. At this spacing, there is little interference between the equipotential shells of individual pores (Fig. 9.7).

When there are n pores per unit leaf area, the resistance of a set of pores r_s can be readily derived from the resistance of the individual pores r_t. For example, if $\delta\chi$ is the difference of water vapour concentration maintained across a set of circular pores with a mean diameter of d, the transpiration rate can be written either as

$$\mathbf{E} = \frac{\delta\chi}{r_s} \quad \text{or as} \quad \frac{n\pi(d^2/4)\delta\chi}{r_t}$$

It follows that

$$r_s = \frac{4(l+\pi d/8)}{\pi n d^2 D}$$

Similar expressions in a variety of units were derived by Penman and Schofield,[104] Heath,[49] and Meidner and Mansfield.[77]

Milthorpe and Penman[79] evaluated the stomatal resistance of wheat leaves with rectangular pores. Refinements in their calculations included: (i) allowing for the stomatal slit getting shorter as the stoma closed; (ii) making the diffusion coefficient a function of the stomatal width to allow for the effect of 'slip' at the stomatal walls. This phenomenon is important when the width is comparable with the mean free path of the diffusing molecules. For example, when the width of the throat was 1 μm, the diffusion coefficient for water vapour was 88% of its value in free air; (iii) making the end-correction for the inner end of the pore 1·5 times the correction for the outer end. (Figure 9.6 suggests that this factor should have been smaller rather than greater than unity.) Figure 9.8 shows the relation between resistance and slit width (a) from figures tabulated by Milthorpe and Penman for wheat stomata, assumed rectangular, and (b) from figures tabulated by Biscoe[10] for sugar beet stomata, assumed elliptical.

Figure 9.8 can be used to estimate the effect of stomatal closure on the *total* resistance to diffusion of water vapour or carbon dioxide for a leaf, taking account of the distribution of stomata over the abaxial and adaxial surfaces. On the assumption that the sugar beet has amphistomatous leaves with the same resistance r_s on each epidermis, the total resistance for each surface is the sum of two resistances in series, i.e. r_s+r_V, and the resistance of the whole leaf is the sum of two resistances in parallel, i.e.

$$\left(\frac{1}{r_V+r_s}+\frac{1}{r_V+r_s}\right)^{-1} = (r_V+r_s)/2$$

When r_V is much smaller than r_s, the leaf resistance is approximately

$r_s/2$. On the assumption that barley has hypostomatous leaves, the resistances of the two surfaces is $r_V + r_s$ (adaxial) and $r_V + x$ (abaxial) where x is a resistance much larger than r_s representing the resistance of the cuticle. Combining these resistances in parallel gives the total leaf resistance as

$$\left(\frac{1}{r_V + r_s} + \frac{1}{r_s + x}\right)^{-1} \simeq r_V + r_s$$

which is approximately equal to r_s when r_V is much smaller than r_s. (Measurements with model bean leaves[137] suggest that the appropriate

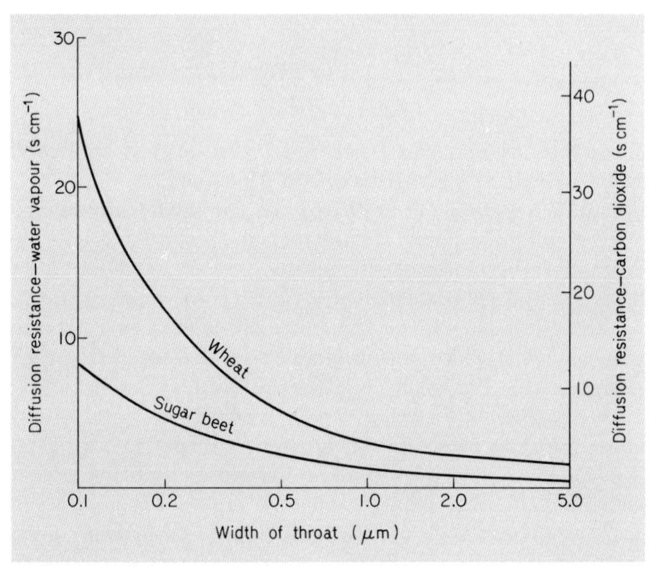

Fig. 9.8 Diffusion resistance calculated for wheat and sugar beet leaves as a function of throat width. Resistances for water vapour and carbon dioxide are given on left- and right-hand axes respectively.

value of r_V for a hypostomatous leaf may be about 30% smaller than the value for a flat plate of the same size because of increased exchange round the edge of the plate.)

In practice, the two surfaces of an amphistomatous leaf often have different stomatal resistances. Relevant equations have been published in the literature but are too cumbersome to reproduce here. As an additional complication, the stomata on the two surfaces may respond in different ways to levels of irradiance and of the water stress in the mesophyll tissue.

The subject of stomatal resistance has been discussed in some detail because stomata play such an important role in governing the exchanges of water vapour by transpiration and of carbon dioxide by photosynthesis. Both physiologists and micrometeorologists are concerned with these exchanges and the concept of stomatal resistance provides a link between two disciplines. Descriptions of stomatal behaviour in response to external and internal stimuli which are beyond the scope of this chapter can be found in most text books of plant physiology.

10

Partitioning of Heat—
(i) Dry systems

The wind and the air and the heat must function in an intermixture throughout the limbs and, although a particular element is bound to be more or less prominent than the others, all must be so intermixed that they are seen to form a unity.

The heat balance of plants and animals will now be examined in the light of the principles and processes considered in previous chapters. The First Law of Thermodynamics states that when a balance sheet is drawn up for the flow of heat in any physical or biological system, income and expenditure must be exactly equal. In micrometeorology, radiation and metabolism are the main sources of income; radiation, convection and evaporation are methods of expenditure. Changes in the amount of heat stored in a system can be determined from its heat capacity and the rate of change of temperature.

The heat balance of a plant or animal can be compared to the water balance of a tank supplied with water from a tap and fitted with an outlet near its base, a system which has provided many problems for elementary arithmetic books (Fig. 10.1). When water is poured into the tank at a steady state, the level rises until the rate of outflow is exactly equal to the supply. If the supply is increased or if the outflow is somewhat restricted, the water level rises again until a new equilibrium is reached. In the same way, the temperature at any point within or on the surface of an organism responds to changes in the supply or in the dissipation of heat. When one or more components of the heat balance change, there is a compensating change of temperature to restore thermal equilibrium, and, when a new

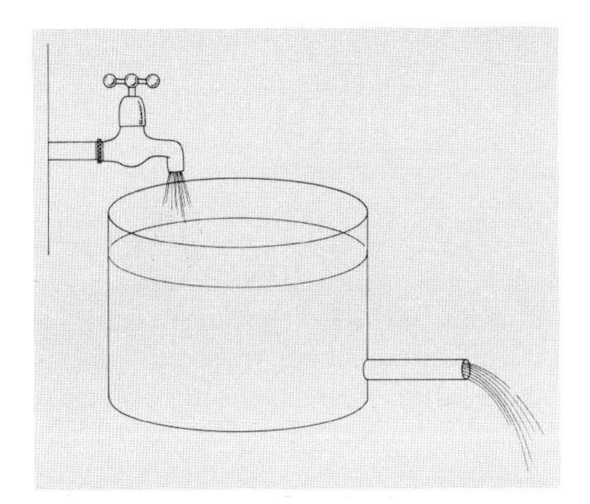

Fig. 10.1 An elementary hydrodynamic analogue of temperature and heat balance.

equilibrium is reached, temperatures become constant throughout the system.

The temperature of plant tissue and of cold-blooded or '**poikilothermic**' animals is dictated solely by the thermal environment and always responds to changes in that environment. The deep body temperature of warm-blooded animals or '**homeotherms**' is maintained nearly constant by physiological mechanisms which alter the rate at which heat is produced by metabolism and dissipated by convection and evaporation. The temperature sensor controlling these mechanisms is equivalent to a constant pressure device in the outlet from a tank, maintaining a fixed water level by altering either the supply or the loss of water.

HEAT BALANCE EQUATION

The heat balance of any organism can be expressed by an equation of the form

$$\bar{\mathbf{R}}+\bar{\mathbf{M}} = \bar{\mathbf{C}}+\overline{\lambda\mathbf{E}}+\bar{\mathbf{J}}+\bar{\mathbf{G}}$$

where the bars indicate that each term is an average heat flux per unit surface area. In this context, it is convenient to define surface area as the area from which heat is lost by convection although this is not necessarily identical to the area from which heat is gained or lost by radiation. The individual terms are

$\bar{\mathbf{R}}$ = net gain of heat from radiation
$\bar{\mathbf{M}}$ = net gain of heat from metabolism

$\overline{\mathbf{C}}$ = loss of sensible heat by convection
$\overline{\lambda\mathbf{E}}$ = loss of latent heat by evaporation
$\overline{\mathbf{J}}$ = rate of change of stored heat
$\overline{\mathbf{G}}$ = loss of heat by conduction to environment

The conduction term $\overline{\mathbf{G}}$ is included for completeness but is negligible for plants and has rarely been measured for animals.

The grouping of terms in the heat balance equation is dictated by the arbitrary sign convention that fluxes directed away from a surface are positive. (When temperature decreases with distance from a surface so that $\partial T/\partial z < 0$, the outward flux of heat $\mathbf{C} = -k(\partial T/\partial z)$ is a positive quantity.) The sensible and latent heat fluxes $\mathbf{C}$ and $\lambda\mathbf{E}$ are therefore taken as positive when they represent losses of heat and as negative when they represent gains. On the left-hand side of the equation **R** and **M** are positive when they represent gains and negative when they represent losses of heat. When both sides of a heat balance equation are positive, the equation is a statement of how the total amount of heat available from sources is partitioned between individual sinks. When both sides are negative, the equation shows how the total demand for heat from sinks is partitioned between available sources.

To analyse the partitioning of heat in ecological problems, it is convenient to distinguish the behaviour of 'dry' and 'wet' systems. In the 'dry' systems which will be analysed in this chapter, the latent heat of evaporation is either much smaller than other fluxes of heat or can be treated as independent of them. The next chapter will consider 'wet' systems in which latent heat is a dominant term in the heat balance equation and is closely related to the sensible heat flux.

To explore the main features of partitioning in dry systems, the relative sizes of heat balance components will be considered first.

Metabolism and radiation

For leaves exposed to bright sunshine $\mathbf{R}_n$ may be 300 to 500 W m^{-2}. and **M** is a negative term representing the net storage of heat in photosynthesis. The maximum rate of net photosynthesis is usually between 2 and 5 g CO_2 m^{-2} h^{-1} depending on species. The assimilation of 1 g of CO_2 produces 0·7 g of carbohydrate with a heat of combustion of about 17 J g^{-1} so the equivalent range of **M** is -7 to -16 W m^{-2}. Because $\mathbf{M}/\mathbf{R}_n$ is small for a sunlit leaf, **M** is comparable with the error in measuring other terms of the heat balance equation and is usually ignored when a thermal account is drawn up for individual leaves or for stands of vegetation. There are a few circumstances however, a combination of low sun and clear sky for example, when the magnitudes of **M** and **R** may be comparable.

During the night, with temperature in the range 10 to 20°C, growing leaves respire at about 2 mg CO_2 m^{-2} h^{-1} per g tissue or 66 mg CO_2 m^{-2} h^{-1} for a leaf with 0·03 m^2 surface area per g dry matter. This is equivalent to a heat source of $\mathbf{M} = +0{\cdot}2$ W m^{-2} so on a clear night when an isolated leaf may lose radiation at a rate $\mathbf{R}_n = -70$ W m^{-2} say, $|\mathbf{M}/\mathbf{R}|$ is only 0·003. On overcast nights $\mathbf{R}_n$ will be an order of magnitude smaller and $\mathbf{M}/\mathbf{R}_n$ an order of magnitude larger. Larger values of $\mathbf{M}/\mathbf{R}_n$ can be obtained on a whole plant basis. The respiration rate of a mature crop stand may reach about 1 g CO_2 m^{-2} field area h^{-1} equivalent to 3 W m^{-2} whereas the radiative loss per unit area of a crop is about the same as the loss per unit area of a horizontal leaf at the top of the crop. In this case $|\mathbf{M}/\mathbf{R}|$ may be about 0·04 on cloudless nights but may approach unity on cloudy nights.

For animals and man, metabolic rate depends on body size, age, activity and diet. Standardized measurements of **basal metabolic rate** are made when subjects have been deprived of food and are resting in a 'thermoneutral' environment in which metabolism is independent of external temperature. The basal metabolism of animals **M** can be related to body weight W by an equation of the form

$$\mathbf{M} = AW^n$$

The index n is close to 0·75 for a wide range of species but the numerical factor A changes between species and decreases with age within species. For interspecific comparisons, Kleiber[58] suggested that A should be taken as 70 kcal day^{-1} per $kg^{0{\cdot}75}$ or 3·4 watts per $kg^{0{\cdot}75}$. The measurements reviewed by Hemmingsen[50] however yield a value of 1·8 watts per $kg^{0{\cdot}75}$ over a very large range of weight from 1 to 10^6 g. During activity, metabolic heat produced by muscles can increase **M** by an order of magnitude (Fig. 10.2). From the measurements reviewed by Hemmingsen, the metabolism of a poikilotherm kept at a body temperature of 20°C is about 5% of the value for a homeotherm of the same weight with a deep body temperature of 39°C. Hibernating mammals metabolize energy at almost the same rate as poikilotherms of equal weight (Fig. 10.3).

Referring metabolic rates to body weight is convenient in nutritional studies but awkward in analyses of heat transfer where fluxes are calculated on an area basis. For any set of objects with different sizes but identical geometry, surface area is proportional to the 2/3 power of volume and therefore to the 2/3 power of weight assuming a constant weight per unit volume. As 2/3 is only slightly less than 0·75, it follows that the basal metabolism of animals with the same body shape should be nearly proportional to their surface area. This proportionality has been demonstrated for mammals ranging in size from a mouse (0·02 kg) to a horse (440 kg). The average basal metabolic rate per square metre of skin is about 50 W m^{-2}. This is much smaller than the net radiant flux absorbed by a dark-coated animal in bright sunshine (say 300 W m^{-2}) but is comparable with the flux that might be absorbed in shade. On a cloudless night, basal metabolism is similar to the net loss of heat by radiation from a surface close to air temperature.

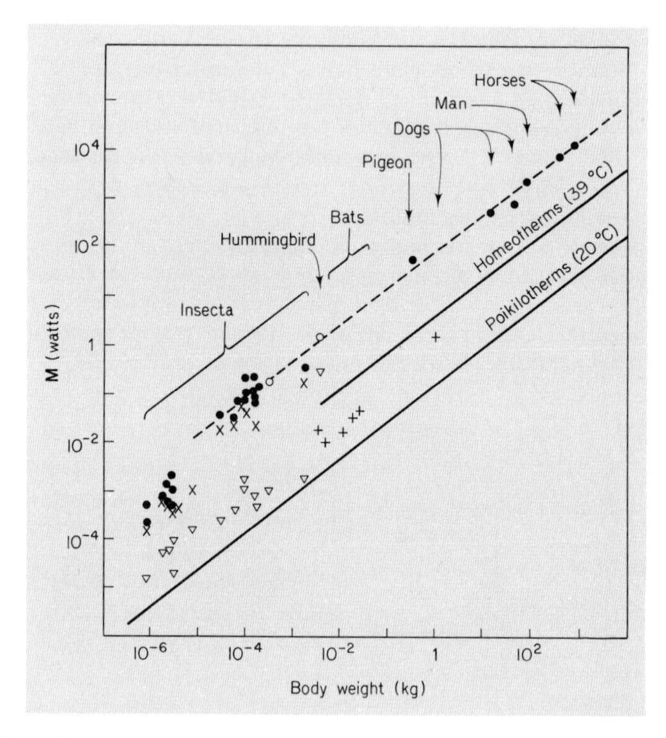

Fig. 10.2 Relation between basal metabolic rate of homeotherms (upper continuous line), maximum metabolism rate for sustained work by homeotherms (upper pecked line) and basal rate for poikilotherms at 20°C (lower continuous line) (from Hemmingsen[50]).

For walking or climbing, the efficiency with which men and domestic animals use additional metabolic energy is about 30%: to work against gravity at a rate of 20 W m^{-2} for example, metabolism must increase by approximately 60 W m^{-2}. The energy expenditure in walking increases with velocity and in man reaches 700 W at 8 km h^{-1}.

For rapid forms of locomotion such as running or flying, the work done

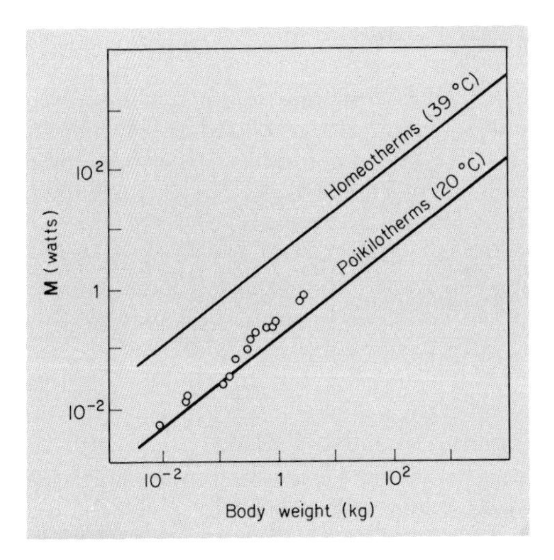

Fig. 10.3 Relation between metabolic rate for hibernating mammals (open circles) and basal metabolic rates for homeotherms and poikilotherms (from Hemmingsen[50]).

against wind resistance is proportional to the wind force times the distance travelled. When the drag coefficient of a moving body is constant, the drag force is proportional to V^2 (p. 82), and as the distance travelled per unit time is proportional to V, the rate of energy dissipation is proportional to V^3. Recent studies on birds reveal a range of airspeeds (e.g. 7 to 12 m s^{-1} for gulls) over which the energy dissipation is almost independent of the velocity of flight revealing a remarkable decrease of drag coefficient with increasing velocity.[142]

Evaporation and metabolism

In the absence of sweat, the loss of latent heat from an animal $\lambda\mathbf{E}$ is usually a small fraction of **M** and takes place both from the lungs as a result of respiration and from the skin as a result of the diffusion of water vapour referred to as 'insensible perspiration'. For a man inhaling completely dry air, the vapour pressure difference between inhaled and expired air is about 52 mbar and the heat used for respiratory evaporation $\lambda\mathbf{E}_r$ is the latent heat equivalent of about 0·8 mg of water vapour per ml of absorbed oxygen.[20] With round figures of 2·4 J mg^{-1} for the latent heat of water and 21 J ml^{-1} oxygen as an appropriate heat of oxidation, $(\lambda\mathbf{E}_r/\mathbf{M})$ is about 10%. When air with a vapour pressure of 12 mbar is breathed, the gradient of vapour concentration between the expired air and the atmosphere decreases to 40 mbar and ($\lambda\mathbf{E}_r/\mathbf{M}$) decreases to about 8%.

The latent heat loss $\lambda\mathbf{E}_s$ from human skin that is not sweating is roughly twice the respiratory loss[20] so that $\lambda\mathbf{E}_s/\mathbf{M} \simeq 0{\cdot}16$ when the vapour pressure is 12 mbar. The total evaporative loss can therefore reach $0{\cdot}24\mathbf{M}$ even when the surface of the skin is dry but the exact figure depends both on atmospheric conditions and on activity.

Sheep and cattle lack glands of the type that allow man to sweat profusely but some species are able to lose a limited amount of water by sweating and others compensate for their inability to sweat by panting in very hot environments. The respiratory system of ruminants can account for 30% of the total evaporative heat loss and the remaining 70% can be ascribed to the evaporation of sweat from the skin surface and from wetted hair. The maximum evaporative heat loss is only a little larger than metabolic heat production whereas a sweating man can lose far more heat by respiration than he produces metabolically.

Interspecific differences in $\lambda\mathbf{E}/\mathbf{M}$ and in mechanisms of evaporation may play an important part in adaptation to dry environments. Relatively small values of $\lambda\mathbf{E}/\mathbf{M}$ have been reported for a number of desert rodents which appeared to conserve water evaporated from the lungs by condensation in the nasal passages where the temperature was about 25°C.[121] The respiratory system operates as a form of counter-current heat exchanger. In contrast, measurements of *total* evaporation from a number of reptiles yield figures ranging from 4 to 9 mg of water per ml of oxygen. These figures imply that nearly all the heat generated by metabolism was dissipated by the evaporation of water, lost mainly through the cuticle, so that $\mathbf{M}-\lambda\mathbf{E}$ was approximately zero.

For different animal species the amount of body water available for evaporation depends on body volume whereas the maximum rate of evaporation depends on surface area. At one extreme, insects have such a large surface/volume ratio that evaporative cooling is an impossible luxury. Man and larger animals can use water for limited periods to dissipate heat during stress and, in general, the larger the animal the longer it can survive without an external water supply.

Convection and long wave radiation

When the surface of an organism loses heat by convection, the rate of loss per unit area is determined by the geometry and scale of the system as well as by windspeed and temperature gradients. Convection is usually accompanied by an exchange of long wave radiation between the organism and its environment at a rate which depends on geometry and on differences of radiative temperature but is independent of scale. The significance of scale can be demonstrated by comparing convective and radiative losses from an object such as a cylinder with diameter d and uniform surface temperature T_s exposed in a wind tunnel whose internal walls are kept at the temperature T of the air flowing through the tunnel with velocity u. When Re exceeds 10^3, the resistance to heat transfer by convection increases with $d^{0.4}$ according to the relation $r_H = d/(\kappa\ \mathrm{Nu}) \propto d^{0.4}V^{-0.6}$ (see Table A.5). The corresponding resistance to radiative

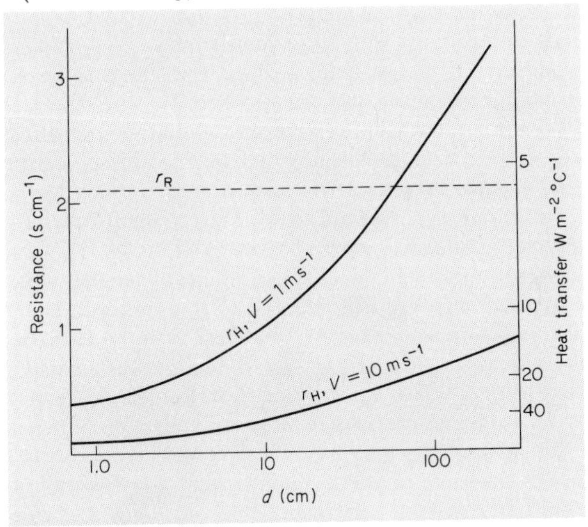

Fig. 10.4 Dependence of aerodynamic resistance r_H and resistance to radiative heat exchange r_R as a function of body size represented by the characteristic dimension of a cylinder d.

transfer r_R can be derived by writing the radiative exchange per unit area as

$$\sigma(T_s^4 - T^4) \simeq 4\sigma T^3(T_s - T) = \rho c_p(T_s - T)/r_R$$

where the resistance $r_R = \rho c_p/4\sigma T^3$ is independent of d. Figure 10.4 compares r_H and r_R for cylinders of different diameters at windspeeds of 1 and 10 m s^{-1} chosen to represent outdoor conditions. Corresponding losses of heat are shown on the right-hand axis for $T_s - T = 1$ K.

Because r_H is similar for planes, cylinders and spheres provided the appropriate dimension is used to calculate Nusselt numbers, a number of generalizations may be based on Fig. 10.4. For organisms on the scale of a small insect or leaf ($0{\cdot}1 < d < 1$ cm), r_H is much smaller than r_R implying that convection is a much more effective mechanism of heat transfer than long wave radiation. The organism is tightly coupled to air temperature but not to the radiative temperature of the environment. For organisms on the scale of a farm animal or a man ($10 < d < 100$ cm), r_H and r_R are of comparable importance at low windspeeds. For very large mammals ($d > 100$ cm), r_H can exceed r_R at low windspeeds and in this state the surface temperature will be coupled more closely to the radiative temperature of the environment than to the air temperature. These predictions are consistent with measurements on locusts and on piglets, for example, and they emphasize the importance of wall temperature as distinct from air temperature in determining the thermal balance of large farm animals in buildings with little ventilation.

The significance of a resistance to heat transfer that decreases with body size has sometimes been overlooked by ecologists. Schmidt–Nielsen[121] regarded both convective and radiative exchanges as proportional to surface area and concluded that in desert conditions 'the small animal with its larger relative surface is in a much less favourable position for maintaining a tolerably low body temperature'. It is true that small animals are unable to keep cool by evaporating body water but when convection is the dominant mechanism of heat loss they can lose heat more rapidly than large animals exposed to the same windspeed. In this context, the main disadvantage of smallness is microclimatic: because windspeed increases with height above the ground, small mammals moving close to the surface are exposed to slower windspeeds than larger mammals. As r_H is approximately proportional to $(d/V)^{0.5}$, an animal with $d = 5$ cm exposed to a wind of 10 cm s^{-1} will be coupled to air temperature in the same way as a much larger animal with $d = 50$ cm exposed to wind at 1 m s^{-1}. Insects and birds in flight and tree-climbing animals escape this limitation.

Heat storage and radiation

The heat balance equation contains a storage term **J** representing quantities such as the change in heat content of a leaf or animal or the heat flux into the soil beneath vegetation.

The size and significance of **J** for an organism can be determined by writing

$$\mathbf{J} = \partial(\mathscr{C}T)/\partial t$$

where $\mathscr{C}$ is the heat capacity in units of J °C^{-1} per m^2 of surface.

Leaves

On the assumption that leaf material has the same specific heat as water, a leaf 1 mm thick will have a heat capacity of 4·2 kJ m^{-2} °C^{-1}. If leaf temperature increased by 5°C per hour on a clear summer morning, **J** would be $(4200 \times 5)/3600 = 6$ W m^{-2}, about two orders of magnitude smaller than the contemporary radiant flux from the sun. Larger values of **J** up to about 100 W m^{-2} can be reached when the sun is appearing or disappearing behind clouds and must be taken into account when energy balances are struck over periods of a few minutes. Over whole days, **J** is always a small term in the energy balance, usually a few per cent of the corresponding net radiation even in forest stands.

Animals

The deep body temperature of most homeotherms is constant enough for diurnal changes of heat storage to be neglected in relation to other components of the heat balance. Some desert animals, however, allow their temperatures to increase by several degrees during the day and it has been suggested that this behaviour may represent an adaptation to conditions of heat stress helping to conserve water. For example, the temperature of dehydrated camels can fluctuate between a minimum of 34°C in the early morning and a maximum of 40°C in the afternoon.[121] For a body weight of 500 kg and a specific heat of 3·4 J g^{-1}, the storage of heat over twelve hours is 10^7 J. If the surface area of a 500 kg camel is assumed to be 6 m^2, the equivalent heat flux is 40 W m^{-2}, about 10 to 20% of maximum net radiation in a desert environment. The advantage of this storage is probably greatest early in the day when air and body temperatures increase rapidly and in this period **J** may exceed 100 W m^{-2}. In the early afternoon, however, air and body temperatures reach a maximum, **J** decreases to zero and then changes sign so that it represents a source of energy. From this point until the net radiation becomes negative the animal is forced to dissipate an additional amount of heat, partly by the evaporation of water. The thermal advantages of an oscillating temperature become more doubtful when the increase of metabolic rate with deep body temperature is taken into account. If the average metabolic rate is 80 W m^{-2} with a Q_{10} of 2, a 6°C increase of temperature over 12 hours will increase metabolic rate to a maximum of 120 W m^{-2} and the average metabolism over the warming period will probably exceed 100 W m^{-2}. This figure implies that at least half the apparent benefit of heat storage (40 W m^{-2}) may be lost by increased metabolism (100 − 80 = 20 W m^{-2}).

Changes of body heat storage are possible even when deep body temperature stays constant, provided there are substantial changes in the average temperature of peripheral tissue and of appendages such as legs and arms. For a man weighing 63 kg, for example, total heat content is estimated to decrease by 8×10^5 J when the temperature of the environment decreases from 30 to 0°C. For a surface area of 1·8 m^2 and a change in air temperature of 5°C h^{-1}, the corresponding release of stored heat is

20 W m^{-2} which is of the same order as the corresponding figure for a camel.

Although changes in the body heat storage of mammals may not appear very significant in relation to radiative fluxes during the day, they are comparable with the loss of heat by long wave radiation on a clear night, often between 50 and 100 W m^{-2} (p. 36). In the two cases just discussed, the heat available from body cooling at night could represent a substantial contribution towards the nocturnal heat budget, conserving metabolic energy during the hours of darkness. By absorbing energy from the environment during the day and releasing it to the environment at night, an animal could reduce the intake of food or the oxidation of body fats. In the desert and in some other environments, the conservation of limited supplies of food may be just as important as the conservation of water.

Soil

The relation between the amount of heat stored in the soil and the heat balance of the surface depends in a complex way on a large number of physical factors. Relevant orders of magnitude can be derived by noting from the relations on p. 131 that the flow of heat into a medium depends on the quantities k/D and ρcD both of which are proportional to $\rho c\sqrt{\kappa}$, a term often called the 'conductive capacity' of a medium.[109] For a prescribed temperature regime at the soil-air interface, the relation between the flow of heat into the soil (conduction) and the flow of heat into the atmosphere (convection) depends on the ratios of the conductive capacities of soil and air. A moist sandy or clay soil will have a conductive capacity of about $1{\cdot}7 \times 10^4$ J m^{-2} $s^{-1/2}$ °C^{-1} and, assuming a representative value of $K_H = 10$ m^2 s^{-1}, the corresponding value for the lower atmosphere is 4×10^3 J m^{-2} $s^{-1/2}$ °C^{-1}. On this basis, rates of heat transfer by conduction and convection should be similar in size. Estimates of the ratio of **G** to **C** must take account of diurnal changes of surface temperature, and the dependence of K_H on height, windspeed and surface

Table 10.1 Ratio of $+\mathbf{G}/+\mathbf{C}$ for daily cycle

u_* m s^{-1}	z_0 cm		
	0·1	1·0	10·0
0·2	1·89	1·55	1·21
0·4	1·00	0·83	0·66
0·6	0·68	0·57	0·46

roughness (see p. 91). Table 10.1 contains estimates by Sellers[124] of $+\mathbf{G}/+\mathbf{C}$ for a range of values of friction velocity u_* and roughness length z_0. The symbol $+\mathbf{G}$ represents the total amount of heat stored in a soil with $\rho c\sqrt{\kappa'} = 1{\cdot}7 \times 10^4$ units during the half cycle when heat is conducted downwards from the surface and $+\mathbf{H}$ is the total heat transfer to the atmosphere when heat is being convected upwards. For the special case of a dry soil, the heat balance equation can be written in the simple form $\mathbf{R} = \mathbf{G} + \mathbf{C}$ so that $\mathbf{G}/\mathbf{R}$ is $\mathbf{G}/(\mathbf{G}+\mathbf{C}) = (1+\mathbf{C}/\mathbf{G})^{-1}$. For the range of values of $\mathbf{G}/\mathbf{C}$ given in the table, $\mathbf{G}/\mathbf{R}$ lies between 0·31 and 0·65.

When a soil is covered with vegetation however, $\mathbf{G}/\mathbf{R}$ is usually about 0·05 to 0·1 during the day, partly because a large fraction of radiant energy is lost by evaporation and partly because the soil surface is shaded by the foliage. At night, when there is no evaporation, $\mathbf{G}/\mathbf{R}$ increases as windspeed decreases (see Table 10.1) and approaches unity on very calm nights when atmospheric turbulence and vertical heat transfer are inhibited by strong inversions. Under these circumstances almost all the heat lost by radiation from a bare soil surface or from vegetation growing on it is supplied by conduction from the underlying soil.

Conduction

Few attempts have been made to measure the conduction of heat from an animal to the surface on which it is lying but Mount[95] has measured the heat lost by young pigs to different types of floor material and has found that the rate of conduction is strongly affected by posture as well as by the difference between deep body temperature and the temperature of the substrate. When the temperature of the floor (and air) was low, the animals assumed a tense posture and supported their trunks off the floor, but as the temperature was raised they relaxed and stretched to increase their contact with the floor. Figure 10.5 shows the heat loss per unit area recalculated from Mount's data for new born pigs in a relaxed position. As the heat flow is approximately proportional to the temperature difference, the thermal resistance equivalent to each type of floor can be calculated from the temperature difference per unit heat flux. The resistances are approximately 8, 17 and 58 s cm^{-1} for concrete, wood and polystyrene respectively. As the corresponding resistances for convective and radiative heat transfer are usually of the order of 1 to 2 s cm^{-1} (Fig. 10.4), it follows that heat losses by conduction will be significant but small when the animals are lying on a material such as concrete with a high thermal conductivity but will be trivial on materials such as wood and polystyrene.

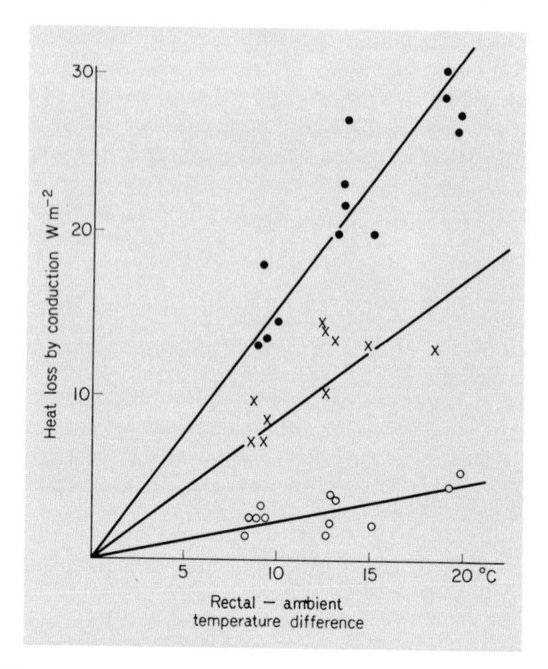

Fig. 10.5 Measurements of heat lost by conduction from a pig to different types of floor covering expressed as watts per square metre of total body area (after Mount[95]), ●, Concrete, ×, wood, ○, polystyrene.

GRAPHICAL ANALYSIS

Graphical methods are useful for solving the heat balance equation of natural surfaces. Porter and Gates[107] have recently described a method for animals regarded as dry systems in the sense that evaporation was assigned fixed values for each species considered. An alternative method will be developed here with the intention of extending its application to 'wet' systems in the next chapter.

If $\boldsymbol{\lambda E}$ is a small term compared with $\mathbf{M}$, the heat balance equation is written in the form

$$\mathbf{R}_n + (\mathbf{M} - \boldsymbol{\lambda}\mathbf{E}) = \mathbf{C} \tag{10.1}$$

where each term applies to unit surface area and bars are omitted for convenience. Provided the emissivity of the surface is unity, the net radiation can be written as

$$\mathbf{R}_n = \mathbf{S}_n + \mathbf{L}_a - \sigma T_0^{\,4} \tag{10.2}$$

where $\mathbf{S}_n$ is the net flux of solar radiation received per unit area of surface, $\mathbf{L}_a$ is the flux of absorbed long wave radiation and T_0 is a mean surface temperature. It is convenient to define a new quantity $\mathbf{R}_{ni}$ as the net radiation that would be received in the same environment by an identical surface at air temperature. The subscript i denotes an isothermal system. Then

$$\begin{aligned}\mathbf{R}_{ni} &= \mathbf{S}_n + \mathbf{L}_a - \sigma T_a{}^4 \\ &= \mathbf{R}_n + \sigma(T_0{}^4 - T_a{}^4) \qquad 10.3\end{aligned}$$

If the value of $\mathbf{R}_n$ given by equation 10.3 is substituted in 10.1, the heat balance equation becomes

$$\begin{aligned}\mathbf{R}_{ni} + (\mathbf{M} - \lambda\mathbf{E}) &= \mathbf{C} + \sigma(T_0{}^4 - T_a{}^4) \\ &= \frac{\rho c_p(T_0 - T_a)}{r_H} + \frac{\rho c_p(T_0 - T_a)}{r_R} \qquad 10.4\end{aligned}$$

The two terms on the right-hand side of the equation represent two simultaneous losses of heat; a real heat loss by convection from a surface at T_0 to surrounding air at T_a; and an apparent heat loss by radiation from a surface at T_0 to a surrounding surface at T_a. The corresponding resistances to heat transfer can be regarded as wired in parallel to give a total resistance r_{HR} such that

$$\frac{1}{r_{HR}} = \frac{1}{r_H} + \frac{1}{r_R} \qquad 10.5$$

At 20°C r_R is about 2·1 s cm^{-1} and when r_H is 0·4 s cm^{-1} or less, r_{HR} is only slightly less than r_H. This derivation of r_{HR} contains the implicit assumption that the areas for convective and radiative exchange are equal. A difference between these areas could readily be allowed for in a more rigorous treatment.

Equations 10.4 and 10.5 can now be combined to give

$$\mathbf{R}_{ni} + (\mathbf{M} - \lambda\mathbf{E}) = \frac{\rho c_p(T_0 - T_a)}{r_{HR}} \qquad 10.4a$$

As a final step to simplify the graphical presentation of heat balances, the isothermal net radiation $\mathbf{R}_{ni}$ is removed from the equation by writing

$$(\mathbf{M} - \lambda\mathbf{E}) = \frac{\rho c_p(T_0 - T_e)}{r_{HR}} \qquad 10.6$$

where T_e is an effective temperature for the environment. The elimination of $(\mathbf{M} - \lambda\mathbf{E})$ from equations 10.4a and 10.6 gives

$$T_e = T_a + \{\mathbf{R}_{ni} r_{HR} / \rho c_p\}$$

and the term in curly brackets is similar to the 'radiation increment' used by Burton and Edholn.[20]

Figure 10.6 provides a graphical method of estimating the temperature increment $T_e - T$ when windspeed and net radiation are known. Given the characteristic dimension of an organism d and the velocity of the surrounding air u, the corresponding value of r_H can be read from the left-hand vertical axis and r_{HR} derived from equation 10.5 can be read from the right-hand axis. From the right-hand section of the figure, the

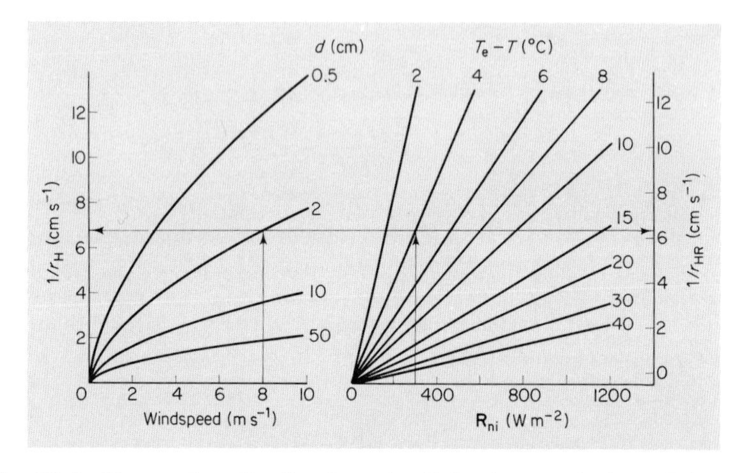

Fig. 10.6 Diagram for estimating thermal radiation increment when windspeed, body size, and net radiation are known. For example, at 8 m s^{-1} an animal with $d = 2$ cm has $r_H^{-1} = 6{\cdot}7$ cm s^{-1} and $r_{HR}^{-1} = 6{\cdot}2$ cm s^{-1}. When $\mathbf{R}_{ni} = 300$ W m^{-2}, $T_e - T$ is 4°C.

coordinates of r_{HR} and $\mathbf{R}_{ni}$ define a unique value of $T_e - T$. An example is shown for $u = 8$ m s^{-1}, $d = 2$ cm, which gives $1/r_H = 6{\cdot}8$ cm s^{-1} from the left-hand axis and $1/r_{HR} = 6{\cdot}3$ cm s^{-1} from the right-hand axis. At $\mathbf{R} = 300$ W m^{-2}, $T_e - T = 4$°C.

In a system where the resistances are fixed, the relation between temperature gradients and heat fluxes can be displayed by plotting temperature against flux as in Fig. 10.7. By definition, a resistance is proportional to a temperature difference divided by a flux, and is therefore represented by the slope of a line in the figure. From a start at the bottom left-hand corner, T_e is determined by drawing a line (1) with slope $r_{HR}/\rho c_p$ to intercept the line $x = \mathbf{R}_{ni}$ at $y = T_e$. The equation of this line is $T_e - T = r_{HR}\mathbf{R}_{ni}/\rho c_p$.

The temperature of the surface T_0 is now found by drawing a second line (2) with the same slope as (1) to intersect $x=(\mathbf{M}-\lambda\mathbf{E})$ at $y=T_0$. The equation of this line is

$$T_0 - T_e = r_{HR}(\mathbf{M}-\lambda\mathbf{E})/\rho c_p$$

(cf. equation 10.6).

Finally, for an animal covered with a layer of hair, a mean skin temperature can be determined if the mean coat resistance r_c is known. Provided that evaporation is confined to the surface of the skin and the

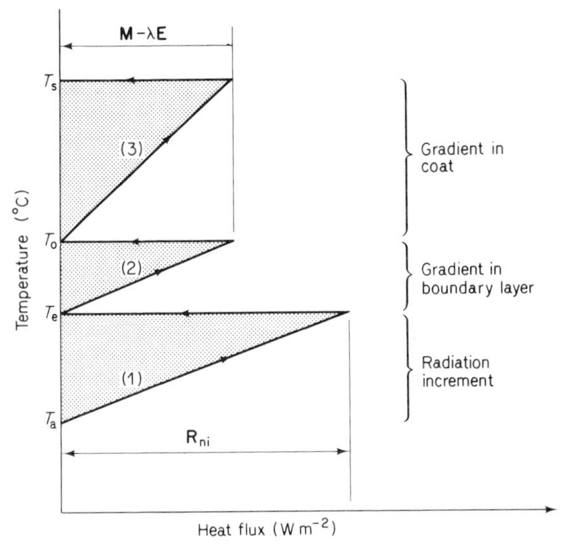

Fig. 10.7 Main features of temperature/heat-flux diagram for dry systems. T_s is skin temperature, T_0 coat surface temperature, T_e effective environment temperature, and T_a air temperature.

respiratory system, the increase of temperature through the coat is represented by the line (3) whose equation is

$$T_s - T_0 = r_c(\mathbf{M}-\lambda\mathbf{E})/\rho c_p$$

This form of analysis can be used to solve two types of problems. When the environment of an animal is prescribed in terms of windspeed, temperature and net radiation, it is possible to estimate a range of physiological states in which the animal can survive in thermal equilibrium with the environment. Conversely, when physiological conditions are specified, a corresponding range of environmental conditions can be established, forming what some writers describe as an ecological 'niche'. Examples now given for a locust, a lizard, a sheep and a man are based on case studies reported in the literature.

CASE STUDIES

Locust

In a monograph describing the behaviour of the Red Locust (*Nomadacris septemfasciata*), Rainey, Waloff and Burnett[111] derive radiation and heat balances for an insect of average size basking on the ground and flying. Both states are represented in Fig. 10.8. At the bottom of the

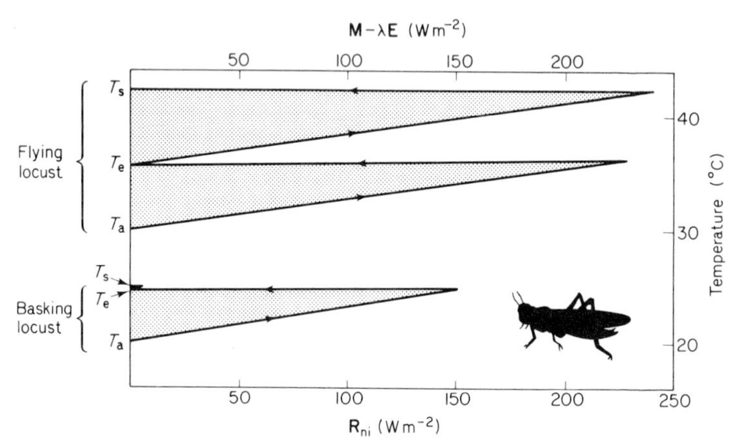

Fig. 10.8 Temperature/heat-flux diagram for locust basking (lower section of graph) and flying (upper section).

diagram, the basking locust is exposed to an air temperature of 20°C but the effective temperature of the environment is about 25°C ($\mathbf{R}_{ni}=150$ W m^{-2}, $r_{HR}=0{\cdot}38$ s cm^{-1}). Because $\mathbf{M}-\boldsymbol{\lambda}\mathbf{E}$ is trivial, the surface temperature is only a fraction of a degree above T_e. The flying locust is exposed to a larger radiant flux density (230 W m^{-2}) but with a relative windspeed of 5 m s^{-1}, the value of r_{HR} is 0·33 s cm^{-1}, somewhat smaller than for the basking locust. The increment T_e-T is therefore only slightly larger than for the basking locust. The difference between body surface and effective temperature is relatively large for the flying locust because the rate of metabolism is much larger than the basal rate and from measurements of oxygen consumption in the laboratory is estimated to be 270 W m^{-2}. The graph predicts that a flying locust should be about 12°C hotter than the surrounding air, consistent with maximum temperature excesses measured by sticking hypodermic thermocouple needles into locusts captured in the field.

Lizards

A very detailed analysis of the heat balance of a lizard (*Sceloporus occidentalis*) was attempted by Bartlett and Gates[5] and Fig. 10.9 summarizes one aspect of their work. Solar elevation is assumed to be 45°C. For a lizard clinging vertically to a tree trunk, the estimated net radiation $\mathbf{R}_{ni}$ ranges from a maximum of about 150 W m^{-2} on the side of the tree facing the sun to 20 W m^{-2} on the shaded side. Lines representing maximum and minimum values of r_{HR} are drawn for air temperatures of

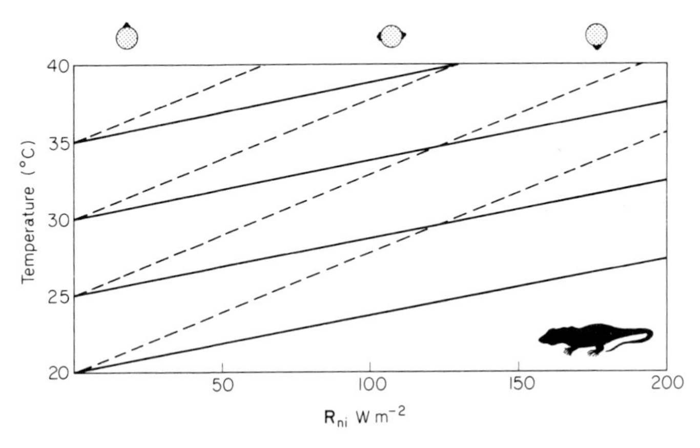

Fig. 10.9 Temperature/heat-flux diagram for lizards on a tree. The sun is due south and the relation between position and irradiance is shown at the top of the diagram. A pair of lines is associated with each air temperature representing minimum resistance to heat transfer (continuous line) and maximum resistance (pecked line).

20, 25, 30 and 35°C. The maximum value corresponds to wind incident on the opposite side of the tree from the lizard, e.g. a north wind when the locust is on the south side: the minimum value corresponds to a position of maximum ventilation, e.g. a lizard on the east or west side of the tree in a north or south wind. If the body temperature of a poikilotherm at rest is within 1°C of surface temperature, the graph can be used to estimate the combination of air temperature and radiation needed to achieve a given body temperature. For example lizards with a temperature preferendum in the range 35–37°C would not seek the shady side of the tree unless the air temperature was above 35°C but would be comfortable on the sunlit side of the tree either at an air temperature of 25°C and minimum ventilation or at 30°C with maximum ventilation. In the original paper, similar predictions were in broad agreement with observations of lizard behaviour on oak trees at a site in California.

Sheep

Figure 10.10 shows the heat balance of sheep with fleeces 1, 4 and 8 cm long exposed to
(i) An air temperature of −10°C and a net radiative loss of −50 W m^{-3}.
(ii) An air temperature of 40°C and a radiative gain of 160 W m^{-2}.
In both cases the animal is supposed to behave like a cylinder with a diameter of 50 cm exposed to a wind of 2 m s^{-1}, a special case of the conditions analysed by Priestley.[110]

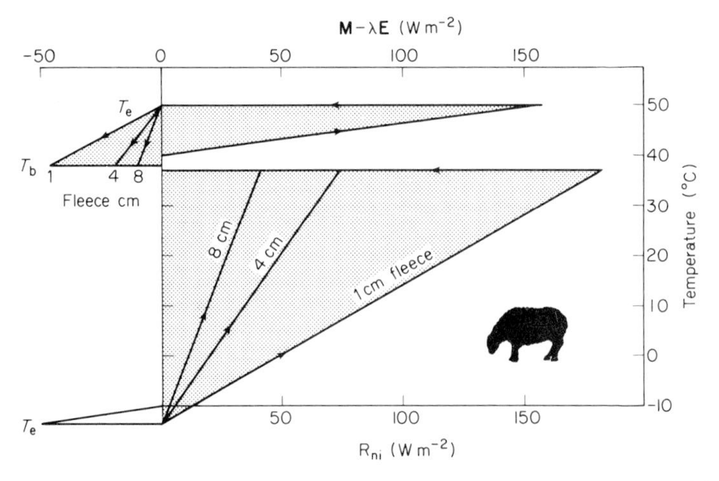

Fig. 10.10 Temperature/heat-flux diagram for sheep with fleece lengths of 1, 4 and 8 cm exposed to air temperatures of −10°C and net radiation of −50 W m^{-2} (lower section); and 40°C with net radiation 160 W m^{-2} (upper section).

For the cold state, T_e is found by drawing a line with slope $r_{HR}/\rho c_p$ to give $T_0 - T_e = 3$°C, at $\mathbf{R}_{ni} = -50$ W m^{-2}. From $T_e = -13$°C, three lines were drawn corresponding to the total resistance $(r_{HR} + r_c)$ for the three values of fleece length taking r_c as 1·5 s cm^{-1} per cm length from Table 8.1. For the heat balance equation in the form

$$T_b - T_e = (\mathbf{M} - \boldsymbol{\lambda}\mathbf{E})(r_{HR} + r_c)/\rho c_p$$

and if $T_b = 37$°C, the lines intercept $T = 37$°C at three values of $\mathbf{M} - \boldsymbol{\lambda}\mathbf{E}$; 42, 74 and 182 W m^{-2}. As the average daily metabolism of a healthy, well-fed sheep is expected to be about 60 or 70 W m^{-2}, the graph implies that a fleece at least 4 cm long is needed to withstand effective temperatures between -10 and -15°C.

For the hot state, the top section of the graph shows that T_e is 50°C in the conditions chosen. Thermal equilibrium cannot be achieved when T_e exceeds T_b unless $\mathbf{M}-\boldsymbol{\lambda}\mathbf{E}$ is negative, i.e. unless more heat is lost by evaporation than is generated by metabolism. $T_b = 38$°C in this case, $\mathbf{M}-\boldsymbol{\lambda}\mathbf{E}$ would need to be -46 W m^{-2} if the fleece length is 1 cm, decreasing to -10 W m^{-2} for an 8 cm fleece, i.e. there would be a flow of sensible heat into the sheep from its environment which would decrease as insulation increased. Studies on sheep in controlled environments show that $\boldsymbol{\lambda}\mathbf{E}$ can reach 90 W m^{-2} under extreme heat stress but, even during a period of minimum activity, **M** is unlikely to be less than 60 W m^{-2}. A figure of -30 W m^{-2} can therefore be taken as a lower limit for $\mathbf{M}-\boldsymbol{\lambda}\mathbf{E}$. The diagram implies that a minimum fleece length of about 2 cm would be needed to withstand the conditions that were chosen to represent heat stress.

Man

The radiation and heat balance of men working in Antarctica was studied by Chrenko and Pugh,[23] and Fig. 10.11 is based on their analysis for a man wearing a black sweater standing facing the sun. The air temperature was only $-7{\cdot}5$°C but because the sun was 22° above the horizon, the radiative load on vertical surfaces facing the sun was exceptionally large. As the wind was light, r_{HR} was relative large, about 1 s cm^{-1}. The top left-hand side of the diagram, referring to the sunlit chest, was constructed from measured temperatures and from the heat flow through the clothing measured with a set of heat flow transducers. The radiation increment was 74°C, the surface of the sweater was at 61°C and the skin was at 38°C. The conduction of heat into the body was $\mathbf{G} = -75$ W m^{-2} and assuming the resistance of the skin over the chest was 0·2 s cm^{-1}, the deep body temperature would be 37°C. The same deep body temperature can be reached starting from the bottom left-hand side of the diagram representing the temperature gradients on the man's back. As the net radiation on the back was only 145 W m^{-2}, T_e was only 6·5°C (cf. 74°C on the chest); the outer surface of the sweater was at 12°C and the skin was at 32°C. There was an outward flow of heat through the skin of $\mathbf{G} = 60$ W m^{-2} and, assuming $T_b = 37$°C, the resistance of the skin is about 1 s cm^{-1}, consistent with the value for vasoconstricted tissue in Table 8.1.

These measurements demonstrate a flow of solar energy through the trunk from chest to back and it would be necessary to integrate this heat flow over the whole body to determine $\overline{\mathbf{M}}-\overline{\boldsymbol{\lambda}\mathbf{E}}$. Such an exercise is

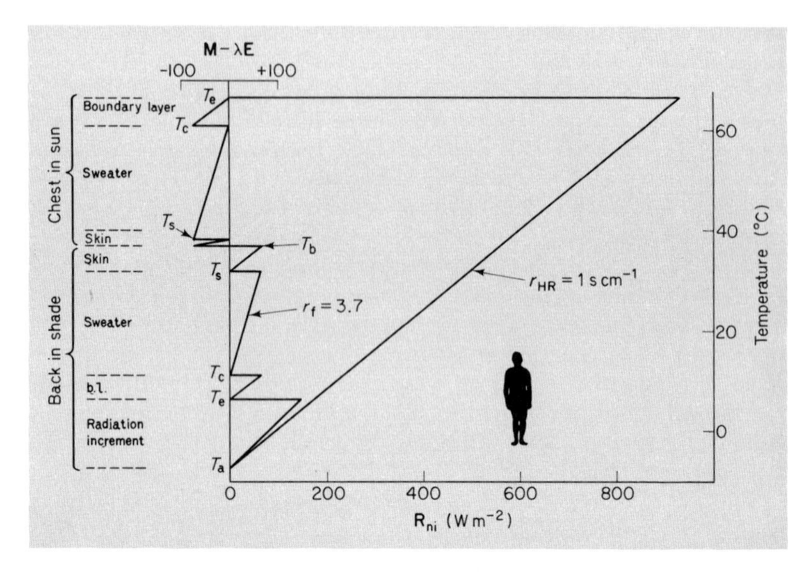

Fig. 10.11 Temperature/heat-flux diagram for a man wearing a black sweater exposed to arctic sunlight and air temperature of −7°C. The lower part of the diagram shows the equivalent temperature and sweater surface temperature on his back (in shade) and the upper part shows the same temperature on his chest (in full sun).

obviously impractical. The heat balance of an animal in a natural environment can be established when $\overline{\mathbf{M}}-\overline{\boldsymbol{\lambda}\mathbf{E}}$ is measured directly or is estimated from relevant laboratory studies but the converse operation of determining $\overline{\mathbf{M}}-\overline{\boldsymbol{\lambda}\mathbf{E}}$ when other terms in the heat balance are known is confined to men or animals in a calorimeter or controlled environment chamber.

II

Partitioning of Heat—
(ii) Wet Systems

A large quantity of water is evaporated by solar heat.

In the examples of partitioning of heat discussed in the last chapter, losses of heat by evaporation were relatively small. Systems in which latent heat is a dominant component of the heat balance will now be analysed with special reference to relations between latent and sensible heat exchanges. Micrometeorological examples include a transpiring leaf and a man covered with sweat. The derivation of wet bulb and equivalent temperatures will be considered first as a basis for constructing a new type of diagram that simplifies the presentation of heat balances in wet systems.

WET BULB AND EQUIVALENT TEMPERATURES

The term 'wet bulb temperature' usually refers to a measurement with a thermometer whose bulb is surrounded by a wet sleeve, but a more fundamental quantity, usually called the **'thermodynamic wet bulb temperature'**, can be derived by considering the behaviour of a sample of air enclosed with a quantity of pure water in a container with perfectly insulating walls. This is an adiabatic system within which the sum of sensible and latent heat remains constant. The initial state of the air can be specified by its temperature T, vapour pressure e and total pressure p. Provided e is smaller than $e_s(T)$, the saturated vapour pressure at T, water will evaporate and both e and p will increase. The increase of latent heat in the system represented by the increase in water vapour concentration must be balanced by a decrease in the amount of sensible heat derived by cooling the air. The process of humidifying and cooling continues until the air becomes saturated at a temperature T', by definition the thermodynamic wet bulb temperature. The corresponding saturated vapour pressure is $e_s(T')$.

To relate T' and $e_s(T')$ to the initial state of the air, the initial water vapour concentration is expressed as $\rho\varepsilon e/(p-e)$ or approximately as $\rho\varepsilon e/p$ when p is much larger than e. When the vapour pressure rises from e to $e_s(T')$, the total change in latent heat content per unit volume is $\lambda\rho\varepsilon[e_s(T')-e]/p$. The corresponding amount of heat supplied by cooling unit volume of air from T to T' is $\rho c_p(T-T')$. (A small change in the heat content of the water vapour is included in more rigorous treatments but is usually unimportant in micrometeorological problems.) Equating latent and sensible heat and rearranging terms:

$$e = e_s(T')-(c_p p/\lambda\varepsilon)(T-T') \qquad 11.1$$

The group of terms $c_p p/\lambda\varepsilon$ is often called the psychrometer constant, is written γ, and has a value of 0·66 mbar °C^{-1} at a temperature of 20°C and a pressure of 1000 mbar (Table A.3, p. 221).

Equation 11.1 can be represented graphically by plotting $e_s(T)$ against

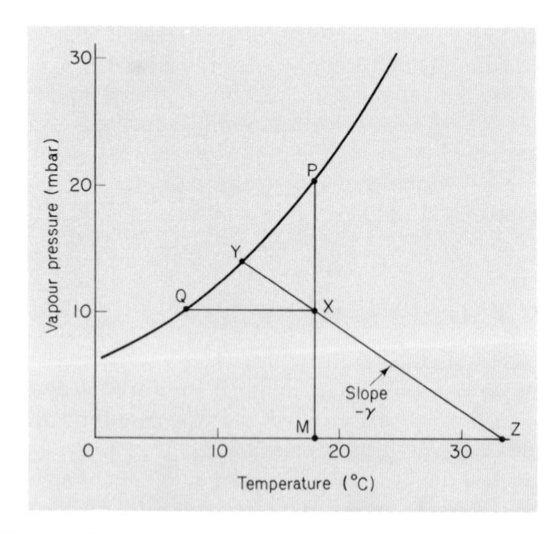

Fig. 11.1 The relation between dry bulb temperature, wet bulb temperature, equivalent temperature, vapour pressure and dew point. The point X represents air at 18°C and 10 mbar vapour pressure. The line YXZ with a slope of $-\gamma$ gives the wet bulb temperature from Y (12°C) and the equivalent temperature from Z (33·3°C). The line QX gives the dew point temperature from Q (7·1°C). The line XP gives the saturation vapour pressure from P (20·6 mbar).

T (Fig. 11.1). The curve QYP represents the relation between saturation vapour pressure and temperature and the point X represents the state of any sample of air in terms of e and T. Suppose the wet bulb temperature of the air is T' and that the point Y represents the state of air saturated at this temperature. The equation of the straight line XY joining the points (T, e), $(T', e_s(T'))$ is

$$e_s(T') - e = \text{slope} \times (T' - T) \qquad 11.2$$

Comparison of equations 11.1 and 11.2 shows that the slope of XY is $-\gamma$. The wet bulb temperature of any sample of air can therefore be obtained by drawing a line with slope $-\gamma$ through the appropriate coordinates T and e to intercept the saturation curve at the appropriate value of T'.

If a sample of air in the state given by X was moved towards Y, the path XY shows how temperature and vapour pressure would change in an adiabatic process, i.e. when the total heat content of the system was constant. Similarly, starting from Y and moving to X, the path YX shows how T and e would change if water vapour were condensed adiabatically from air that was initially saturated. As condensation proceeded, the temperature of the air would rise until all the vapour had condensed. This state is represented by the point Z at which $e = 0$. The corresponding temperature θ is called the **'equivalent temperature'** of the air. As Z has coordinates $(\theta, 0)$, the equation of the line ZX can be written in the form

$$\theta = T + e/\gamma \qquad 11.3$$

Alternatively, the equation of XZ can be written

$$\theta = T' + e_s(T')/\gamma \qquad 11.4$$

showing that the equivalent and wet bulb temperatures are uniquely related. Both T' and θ are conservative (i.e. constant) temperatures when water is evaporated or condensed adiabatically within a sample of air.

Three other ways of specifying the water vapour content of air can be derived from Fig. 11.1 by drawing lines through X parallel to both the axes, intersecting the saturation vapour pressure curve at Q and P. The saturation vapour pressure of the air $e_s(T)$ is given by PM, the relative humidity is XM/PM and the saturation deficit $[e_s(T) - e]$ is PX. The dew point temperature T_d is given by the abscissa of Q.

Temperature enthalpy diagram (TED)

In the adiabatic enclosure considered in the last section, the sum of sensible and latent heat was constant during evaporation and condensation. If this condition is relaxed, a quantity of heat Q may be absorbed by the air in the enclosure when its temperature changes from T_1 to T_2 and its vapour pressure changes from e_1 to e_2. Then the heat balance of the air can be written

heat absorbed = increase of sensible heat + increase of latent heat

or

$$\begin{aligned} Q &= \rho c_p(T_2 - T_1) + \lambda\rho\varepsilon(e_2 - e_1)/p \\ &= \rho c_p(T_2 - T_1) + \rho c_p(e_2 - e_1)/\gamma \\ &= \rho c_p(\theta_2 - \theta_1) \end{aligned} \qquad 11.5$$

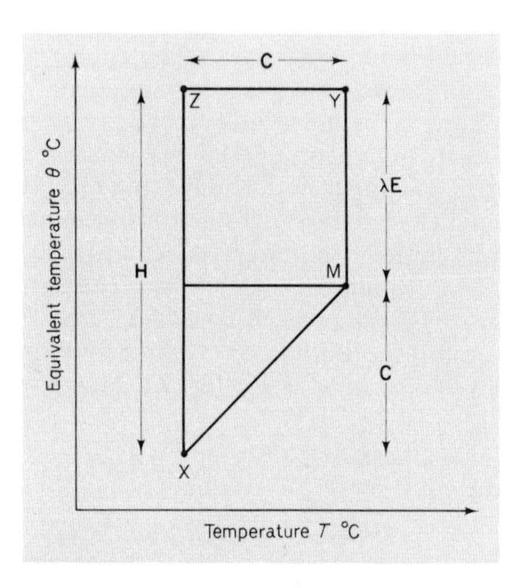

Fig. 11.2 The elements of a temperature-enthalpy diagram (TED). A change of state from X to Y can be achieved by supplying energy **H** represented by XZ which is partitioned into sensible heat **C** (ZY) and latent heat **λE** (YM). The line XM with unit slope is an isobar of vapour pressure.

This equation shows that the equivalent temperature θ is the potential which determines changes of total heat or **enthalpy**, just as the conventional temperature T determines changes of sensible heat. Total and sensible heat exchange can therefore be represented on a temperature enthalpy diagram or TED in which the ordinate θ is a measure of enthalpy and the abscissa is the temperature T (Fig. 11.2). An isothermal process is shown on the TED by a line parallel to the θ axis and an adiabatic process by a line parallel to the T axis. Any change of state from X to Y can be achieved by combining an isothermal process XZ and an adiabatic process ZY. The figure XZYM represents the whole system:

$$\text{change of sensible heat} = \rho c_p(T_2 - T_1) \propto \text{ZY}$$

$$\text{change of latent heat} = \rho c_p \left(\frac{e_2 - e_1}{\gamma}\right) \propto \text{YM}$$

$$\text{change of total heat} = \rho c_p \left[(T_2 - T_1) + \frac{e_2 - e_1}{\gamma}\right] \propto \text{ZX}$$

The heat balance of a wet surface can be analysed in the same way after taking account of the diffusion resistances for heat and vapour transfer. If the state of the surface is given by T_2, e_2, θ_2, and the state of the air by T_1, e_1, θ_1, total heat loss from the surface **H** will be the sum of sensible and latent heat losses or

$$\mathbf{H} = \mathbf{C} + \lambda\mathbf{E}$$

$$= \frac{\rho c_p(T_2 - T_1)}{r_H} + \frac{\rho c_p}{\gamma}\frac{(e_2 - e_1)}{r_V} \qquad 11.6$$

where r_H and r_V are the appropriate diffusion resistances for heat and vapour.

If γ is replaced by $\gamma^* = \gamma(r_V/r_H)$, equation 11.6 can be written

$$\mathbf{H} = \frac{\rho c_p}{r_H}(T_2 - T_1) + \frac{p c_p(e_2 - e_1)}{\gamma^* r_H}$$

$$= \frac{\rho c_p}{r_H}(\theta_2{}^* - \theta_1{}^*) \qquad 11.7$$

where θ^* is an apparent equivalent temperature defined by

$$\theta^* = T + e/\gamma^* \qquad 11.8$$

A TED can now be constructed using γ^* and θ^* in place of γ and θ. Whereas θ is a fundamental property of air, θ^* depends on the physics of evaporation at a particular surface. The device of replacing θ by θ^* allows Fig. 11.2 to be used to describe the heat balance of any surface from which water is evaporating when the relevant physical conditions are known.

SENSIBLE AND LATENT HEAT EXCHANGE

Wet bulb thermometer

When water evaporates from the sleeve covering a wet bulb thermometer, the resistance to vapour diffusion is given by $r_V = d/(D\ \text{Sh})$ where d and Sh are the relevant diameter and Sherwood number and D is the diffusion coefficient of water vapour. When the bulb is surrounded by surfaces at air temperature, the resistance to heat transfer is r_{HR}, derived by combining radiative and convective resistances in parallel (p. 163). Then provided there are no extraneous sources of heat, e.g. from solar radiation or from conduction along the stem of the thermometer, the rate at which heat is lost by evaporation $\rho c_p(e_2 - e_1)/\gamma r_V$ must be exactly equal to the rate at which it is gained by convection and radiation

$$\rho c_p(T_1 - T_2)/r_H$$

The system is represented in a TED by a straight line parallel to the temperature axis and intersecting the saturation line at the temperature of the wet bulb thermometer T_w. The equation of the line is

$$e = e_s(T_w) - \gamma^*(T - T_w) \qquad 11.9$$

which is the psychrometer equation used to find the vapour pressure of air from readings of dry and wet bulb temperature. In standard tables for freely ventilated wet bulbs γ^* is usually assumed equal to $\gamma = 0{\cdot}66$ mbar °C^{-1} but this assumption is strictly correct only in the special case $r_{HR} = r_V$. By putting $r_H = (D/K)^{2/3} r_V$ (p. 136) and $r_R = 2{\cdot}1$ s cm^{-1} (at 20°C), it can be shown that $r_{HR} = r_V$ only when r_H is 0·23 s cm^{-1}. When r_H is larger than this value, r_{HR} is smaller than r_V, γ^* is larger than γ and T_w should be higher than the thermodynamic wet bulb temperature T. Conversely when r_H is smaller than 0·23 s cm^{-1}, T_w should be less than T'.

When r_H and r_R are comparable in size, the value of $\gamma^* = \gamma r_V(1/r_H + 1/r_R)$ is a function of all three resistances. Because r_V and r_H are functions of windspeed and r_R is not, γ^* decreases with increasing ventilation and when r_V is much smaller than r_R, γ^* tends to a constant minimum value independent of windspeed, viz. $\gamma^* = \gamma(r_V/r_H) = \gamma(K/D)^{2/3}$. Equation 11.9 shows that when a psychrometer is exposed in a stream of air whose temperature and vapour pressure are fixed, the wet bulb depression will increase as the ventilation rate increases, reaching a maximum value when γ^* reaches its minimum value. In the Assmann psychrometer often used as a standard instrument for measuring humidity, air is drawn at about 3 m s^{-1} over a bulb with a diameter of 3 mm. The corresponding resistances are $r_V = 0{\cdot}149$ s cm^{-1} and $r_{HR} = 0{\cdot}156$ s cm^{-1} giving $\gamma^* = 0{\cdot}63$ mbar °C^{-1}. With this and similar instruments, the error involved in using 0·66 mbar °C^{-1} instead of γ^* can be ignored in practice.

Wet surface

The rate at which water evaporates from a wet leaf or a free water surface can be determined if the total energy **H** available for sensible and latent heat transfer is assumed equal to $\mathbf{R}_n - \mathbf{G}$, i.e. the difference between the net amount of heat available from radiation and the rate at which heat is stored in the system. The temperature and vapour pressure of the airstream are represented by the point X on a TED constructed with an appropriate value of $\gamma^* = \gamma(r_V/r_H)$. Because θ increases much faster than T, it is convenient to make the scale of T four times larger than the scale of θ. When the TED is scaled in this way, a line of equal vapour pressure *appears* to have a slope of 1 in 4 but has a true slope of unity (Fig. 11.3).

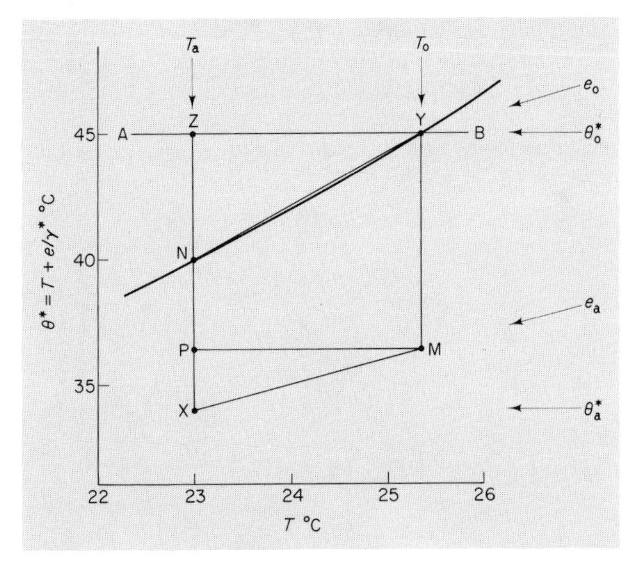

Fig. 11.3 TED for evaporation from a wet surface in a state represented by Y into air in a state represented by X. NY is the saturation vapour pressure curve.

If T_0, e_0, θ_0^* are the temperature, vapour pressure and apparent equivalent temperature of a surface (Y in Fig. 11.3) and T_a, e_a, θ_a^* are the corresponding values in the air X, the heat balance equation is

$$\mathbf{H} = \mathbf{C} + \lambda\mathbf{E}$$

$$= \frac{\rho c_p (T_0 - T_a)}{r_H} + \frac{\rho c_p}{\gamma^*} \frac{(e_0 - e_a)}{r_H}$$

$$= \frac{\rho c_p}{r_H} (\theta_0^* - \theta_a^*) \qquad 11.10$$

The value of θ_0^* can therefore be calculated as $\theta_a^*+(\mathbf{H}r_H/\rho c_p)$ and the line AB represents states with this equivalent temperature. If the air in contact with the surface is saturated, the state of the surface is specified by the coordinates of the point Y where AB intersects the saturation curve. To determine T_0, $\mathbf{C}$ and $\gamma\mathbf{E}$, the line XPNZ is drawn at right angles to AB cutting the saturation curve at N, YM is parallel to ZX where XM is a line with constant vapour pressure e_a and unit slope and ZP=YM. The lines PM and PX therefore represent the same temperature difference T_0-T_a. The slope of the chord YN (almost indistinguishable from the curve) is $\partial(T+e/\gamma^*)/\partial T$ or $1+(\Delta/\gamma^*)$ where $\Delta=\partial e/\partial T$ is the slope of the saturation vapour pressure curve at a point between N and Y.

The TED can now be interpreted by reference to Fig. 11.2; total heat transfer is proportional to ZX, sensible heat to ZY=PM and latent heat to YM=ZP.

To determine latent heat in terms of known parameters the length of YM is determined as

$$\begin{aligned}\mathrm{YM} &= \mathrm{ZP} = \mathrm{ZN}+\mathrm{NX}-\mathrm{PX}\\ &= \mathrm{ZY}\left(1+\frac{\Delta}{\gamma^*}\right)+\mathrm{NX}-\mathrm{PX}\end{aligned}$$

Inserting coordinates gives

$$\frac{(e_0-e_a)}{\gamma^*} = (T_0-T_a)\left(1+\frac{\Delta}{\gamma^*}\right)+\frac{[e_s(T)-e]}{\gamma^*}-[T_0-T_a]$$

and multiplying throughout by $\rho c_p/r_H$ gives

$$\boldsymbol{\lambda}\mathbf{E} = \frac{\Delta}{\gamma}\,\mathbf{C}+\frac{\rho c_p[e_s(T)-e]}{\gamma^* r_H} \qquad 11.11$$

when $\mathbf{C}$ is substituted for $\rho c_p(T_0-T_a)/r_H$. This equation can be used to find $\boldsymbol{\lambda}\mathbf{E}$ by putting $\mathbf{C}=\mathbf{H}-\boldsymbol{\lambda}\mathbf{E}$ and rearranging terms to get

$$\boldsymbol{\lambda}\mathbf{E} = \frac{\Delta\mathbf{H}+\{\rho c_p[e_s(T)-e]/r_H\}}{\Delta+\gamma^*} \qquad 11.12$$

Alternatively when $\boldsymbol{\lambda}\mathbf{E}$ is set equal to $\mathbf{H}-\mathbf{C}$ in equation 11.11 it follows that

$$\mathbf{C} = \frac{\gamma^*\mathbf{H}-\{\rho c_p[e_s(T)-e]/r_H\}}{\Delta+\gamma^*} \qquad 11.13$$

and the surface temperature T_0 can be found from this expression using

$$T_0 = T_a+\mathbf{C}r_H/\rho c_p \qquad 11.14$$

Equation 11.12 was first derived in a slightly different form by Penman[102] and has been used in many different forms to estimate rates of evaporation from water surfaces and from vegetation. A formally similar equation can be derived in terms of the isothermal net radiation setting $\mathbf{H}$ equal to $\mathbf{R}_{ni}-\mathbf{G}$, replacing r_H by r_{HR} and defining γ^* as $\gamma(r_V/r_{HR})$.

When $\mathbf{H}=0$, the equation gives the rate of evaporation from an adiabatic system such as a well-ventilated wet bulb screened from the sun. When $\mathbf{H}$ is finite, the latent heat loss increases by the amount $\Delta\mathbf{H}/(\Delta+\gamma^*)$ and the sensible heat loss increases by the complementary quantity $\gamma^*\mathbf{H}/(\Delta+\gamma^*)$. The terms Δ and γ^* can therefore be regarded as weighting factors which determine the partitioning of radiant energy between evaporation and convection. When the saturation deficit is zero, the formula predicts the rate of evaporation when $\mathbf{H}>0$ and the rate of condensation when $\mathbf{H}<0$.

To calculate the evaporation from an open water surface for periods of a week or more, Penman estimated net radiation, saturation deficit, temperature and windspeed from climatological data. Measurements of evaporation from a small tank of water were used to derive a linear relation between r_H^{-1} and windspeed at 2 m. For this system, resistances for heat and vapour transfer were assumed identical so γ^* was taken as 0·66 mbar °C^{-1}. The rate of evaporation from a free water surface, estimated on this basis, is often written $\mathbf{E}_0$. Penman found that the evaporation from turf $\mathbf{E}_T$ could be expressed as $f\mathbf{E}_0$ where the factor f increased from 0·6 in winter to 0·8 in summer (at latitude 52°N). Initially, $\mathbf{E}_T$ was regarded as a potential rate of evaporation from short green vegetation completely covering the ground and freely supplied with water. Experience on experimental sites ranging from field plots to large catchments has shown that the restriction to short green cover is unnecessary. In temperate climates, the quantity $f\mathbf{E}_0$ is usually within $\pm 15\%$ of the transpiration from vegetation growing in soil where roots have free access to water, irrespective of species.

Measurements of crop water use in arid climates show that evaporation rates sometimes exceed $\mathbf{E}_T$ substantially. This failure of the formula is usually ascribed to an 'oasis' effect, but it is more likely that the empirical factor $f=\mathbf{E}_T/\mathbf{E}_0$ is larger in arid than in temperate climates. The empirical link between $\mathbf{E}_T$ and $\mathbf{E}_0$ can be removed by introducing a physiological resistance depending on the diffusion of water vapour through the stomata as shown in the next section. When this procedure is followed, the Penman formula can safely be used for irrigated plots in dry climates provided climatological measurements are confined to the boundary layer characteristic of the wet area.[148] The Penman formula has been successfully applied to estimate the evaporation from reservoirs, lakes, catchments and crops in a wide variety of climates throughout the world.

Leaves

The evaporation from a single leaf can be estimated from the Penman formula if **H** is taken as the net radiation absorbed by both sides of the leaf and if temperature, saturation deficit and windspeed are measured in the immediate environment. Figure 11.4 shows an equivalent circuit in which the resistance for heat transfer is r_H for each side of the leaf and $r_H/2$ for the two sides in parallel. The resistance to vapour transfer for

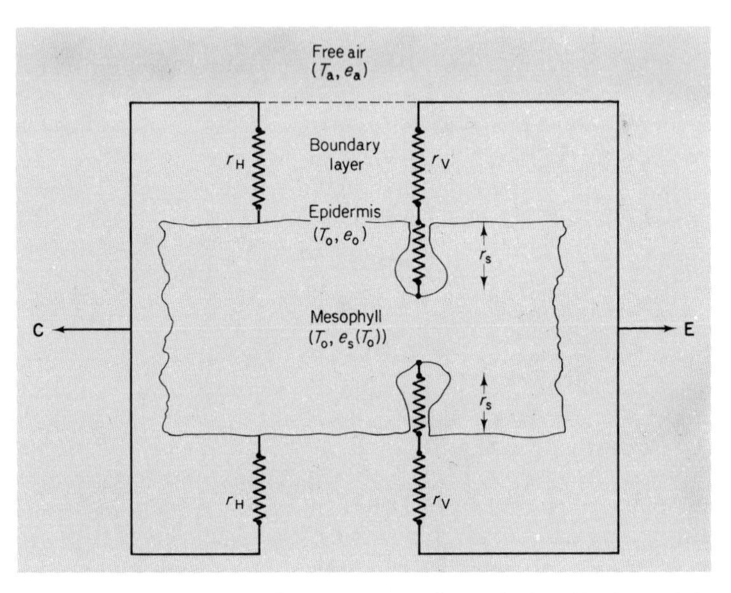

Fig. 11.4 Electrical analogue for transpiration from a leaf and leaf heat balance (see also Fig. 9.3).

each side of the leaf is the sum of a boundary layer resistance r_V and a stomatal resistance r_s. The ratio γ^*/γ therefore assumes values of $(r_V+r_s)\div(r_H/2)$ for a hypostomatous leaf and $(r_V+r_s)/2\div(r_H/2)$ for an amphistomatous leaf with the same stomatal resistance on both surfaces. In general

$$\gamma^* = n\gamma(r_V+r_s)/r_H$$
$$= n\gamma(1+r_s/r_H) \qquad 11.15$$

where $n=1$ (amphistomatous leaf) or $n=2$ (hypostomatous leaf).

Then provided that the cell walls where evaporation occurs are at the same temperature T_0 as the epidermal cells, the heat balance of the leaf can be written

$$\mathbf{H} = \frac{\rho c_p(T_0 - T_a)}{r_H} + \frac{\rho c_p[e_s(T_0) - e_a]}{\gamma^* r_H}$$

which is identical to equation 11.10. It follows that the Penman equation 11.12 can be used to determine the evaporation from a leaf provided γ^* is determined from equation 11.15.[86]

Figure 11.5 shows how the evaporation from a leaf can be derived

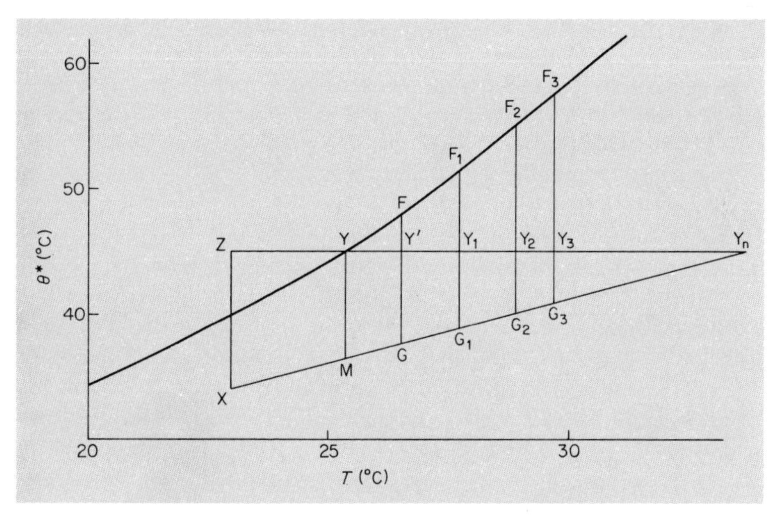

Fig. 11.5 TED for transpiration from a leaf into air in state X. The lines FY′, F_1Y_1, etc. are proportional to the vapour pressure gradient between the intercellular spaces and the epidermis and Y′G, Y_1G_1, etc are proportional to the gradient between the epidermis and the ambient air.

graphically. The figure XZYM is reproduced from Fig. 11.3 on a smaller scale. The presence of stomata does not alter the equivalent temperature of the leaf surface, 45°C in the example shown, but as the air in contact with the leaf surface is not saturated its state cannot be represented by a point on the saturation vapour pressure curve. Suppose the real state of the surface air is represented by Y′, then the isothermal line FY′G is drawn to intersect the saturation vapour pressure curve at F and the vapour pressure line XM at G. The point F represents the state of air in contact with wet cell walls. The ratio FY′/Y′G is the vapour pressure gradient between the cell walls and the leaf surface divided by the gradient between the surface and the surrounding air and this ratio is

r_s/r_a. Thus when r_s/r_a is known, the corresponding point Y′ can be determined. The figure shows Y_1, Y_2 and Y_3 for $r_s/r_a = 1:1$, $2:1$ and $3:1$ respectively. Note that as the stomata close, the temperature of the leaf surface rises until it reaches the point Y_n on XM where r_s/r_a is infinite. Then evaporation is zero and the energy balance is simply $\mathbf{H}=\mathbf{C}$. The rate of evaporation can be derived from Fig. 11.5 by taking YG as $r_a/(r_a+r_s)$ times the evaporation rate and then proceeding with the analysis applied to equation 11.10.

Several implications of the formula will now be considered:

(i) When γ^* is constant, the predicted rate of transpiration is finite when **H** is zero and increases linearly with **H**. In practice, stomatal resistance increases as visible radiation decreases so γ^* gets larger as **H** gets smaller. Using a relation between r_s and radiation measured on barley leaves in the field, Biscoe[10] found that the predicted rate of transpiration was almost proportional to **H**.

(ii) The relation between evaporation rate and windspeed depends on the ratio of sensible to latent heat transfer. By differentiating the equation with respect to r_H, it can be shown that $\boldsymbol{\lambda}\mathbf{E}$ is independent of r_H and hence of windspeed when $\boldsymbol{\lambda}\mathbf{E}/\mathbf{C}=\Delta/n\gamma$.[86] When $\boldsymbol{\lambda}\mathbf{E}/\mathbf{C}$ exceeds this critical value, an increase of windspeed increases the latent heat loss at the expense of the sensible loss in such a way that $\boldsymbol{\lambda}\mathbf{E}+\mathbf{C}$ stays constant. This behaviour is expected intuitively because the evaporation from a free water surface always increases with windspeed. When $\boldsymbol{\lambda}\mathbf{E}/\mathbf{C}$ is smaller than the critical

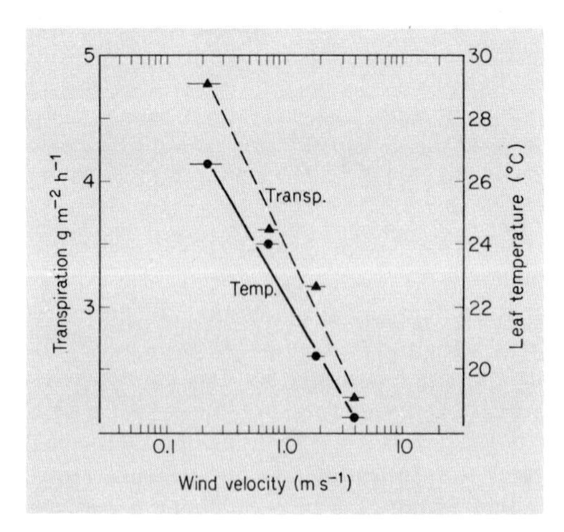

Fig. 11.6 The change of transpiration rate and leaf temperature with windspeed for a *Xanthium* leaf exposed to radiation of 700 W m^{-2} at an air temperature of 15°C and 95% relative humidity (from Mellor *et al.*[78]).

value, however, an increase of windspeed increases **C** at the expense of **λE**: evaporation decreases as windspeed increases. This behaviour has been demonstrated in the laboratory[78] (Fig. 11.6) and inferred from measurements in the field.

(iii) Increasing the stomatal resistance of a leaf decreases the rate of evaporation and increases the sensible heat loss when **H** is constant. Stomatal closure therefore increases the temperature of leaf tissue as shown in Fig. 11.5. Many species have a minimum stomatal resistance of 1–2 s cm^{-1} so in bright sunshine and in a breeze, small leaves are expected to be only 1–2°C hotter than the surrounding air. Greater excess temperatures are observed on very large leaves in a light wind because r_H is large (see Fig. 7.1). From the same set of calculations, it can be shown that the relative humidity of air in contact with the epidermis will usually be similar to the relative humidity of the ambient air and this feature of leaf microclimate may have important implications for the activity of fungi which need a very high relative humidity to reproduce and grow.

Dew

When **H** is negative at night, condensation will occur on a leaf when the numerator of equation 11.12 is negative, i.e. when $-\Delta\mathbf{H}$ exceeds $\rho c_p(e_s(T)-e)/r_H$. The rate of dew formation can be calculated from the

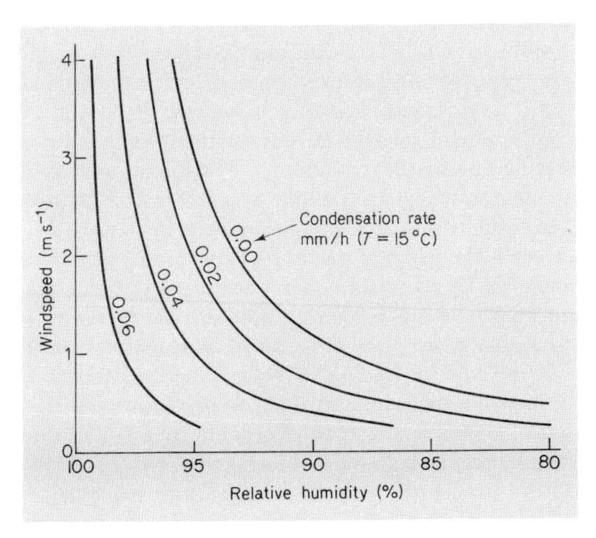

Fig. 11.7 The rate of condensation on a horizontal plane exposed to a cloudless sky at night when the air temperature is 15°C, as a function of windspeed and relative humidity.

formula putting $\gamma^* = \gamma(r_V/r_H)$. When the air is saturated the predicted maximum rate of dew formation on clear nights is about 0·06 to 0·07 mm per hour but may be much less in unsaturated air[84] (Fig. 11.7). These estimates are consistent with the maximum quantities of dew observed on leaves and on artificial surfaces, about 0·2 to 0·4 mm per night depending on site and circumstances. As these quantities are an order of magnitude smaller than potential evaporation rates, dew rarely makes a significant contribution to the balance of vegetation even in arid climates.

Evaporation of sweat

The heat balance of a nude man covered with sweat can be analysed in the same way as the heat balance of a wet leaf and the rate of evaporation can be calculated from the Penman formula when values of the relevant parameters are known. In this case, the net heat **H** will be taken as the sum of the metabolic heat load **M** and the isothermal net radiation $\mathbf{R}_{ni}$ which can be calculated on the assumption that a body intercepts radiation like a cylinder (Chapter 4). The resistance to vapour transfer is $d/(D\ \mathrm{Sh})$ where the characteristic dimension d is often taken as 34 cm for a standard man, and the resistances to heat transfer are $d/(\kappa\ \mathrm{Nu})$ and r_R in parallel. For forced convection in a windspeed of about 2 m s^{-1}, r_H is 1 s cm^{-1} and with $r_R = 2{\cdot}1$ s cm^{-1}, $r_{HR} = 0{\cdot}68$ s cm^{-1}. The corresponding value of r_V is 0·9 s cm^{-1} so $\gamma^* = (r_V/r_{HR})\gamma = 0{\cdot}87$ mbar °C^{-1}. These values of r_V and r_{HR} will be taken as standard in the following discussion.

Just as the rate of transpiration from a leaf cannot be determined accurately until its stomatal resistance is known, the rate at which sweat evaporates from a man in a given environment cannot be determined without knowing how fast it is produced. The maximum rate of sweating that a normal man can sustain is about 1 kg h^{-1} and if his surface area is 1·8 m^2 the equivalent rate of evaporative heat loss is 375 W m^{-2}. The rate of sweating is determined partly by skin temperature and partly by metabolic rate. When a subject reports that a particular combination of clothing and environment is 'comfortable' and there is no visible evidence of sweat, his mean skin temperature is usually about 32 to 33°C. When the heat load on his body is increased, e.g. by exercise or exposure to radiation, or when the temperature of the environment is increased, a small amount of excess heat can be dissipated by a rise in skin temperature involving a faster circulation of blood in the peripheral tissue. When the temperature gradient in this tissue becomes too small to conduct metabolic heat out of the body, sweat is released rapidly from glands a few millimetres below the surface of the skin. Dependent on the nature of the environment and of clothing, sweat may evaporate without appearing at the skin surface (insensible perspiration), or may wet part of the body only, or may cover the whole body. When the rate of evaporation from any area is smaller than the rate at which sweat is produced there, the sweat trickles over the surface of the skin or is lost by dripping, a wasteful process because the water lost from the body plays no part in relieving heat stress.

Temperature gradients for wet bodies

The case of a fully wetted body will be considered first. Laboratory experiments show that many subjects regard the environment as very uncomfortable when skin temperature rises above about 35°C (95°F) and as

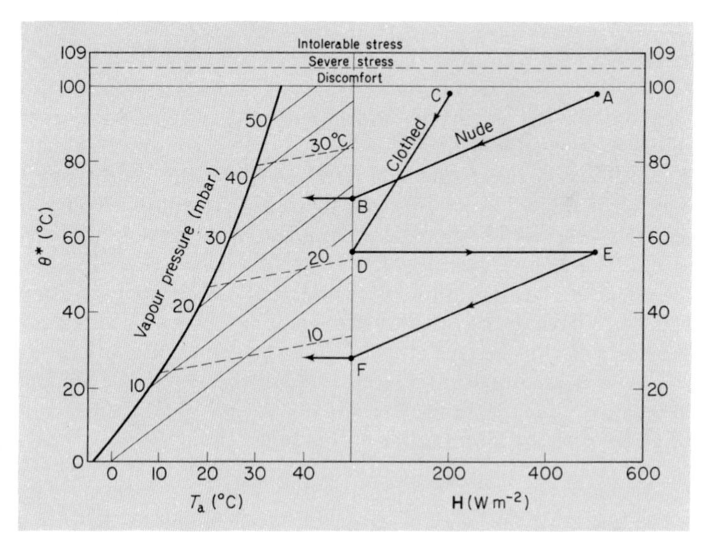

Fig. 11.8 Apparent equivalent temperature and heat-flux diagram for a clothed and a nude man (right-hand section) and the relation between apparent equivalent temperature, vapour pressure and air temperature (left-hand section). The nude man with a total heat load of 560 W m^{-2} can avoid discomfort if the apparent equivalent temperature of the environment is less than the value at B_2 (70°C) but the clothed man with a heat load of only 200 W m^{-2} must stay in an environment with an apparent equivalent temperature less than 28°C (point F). The pecked lines are isotherms of wet bulb temperature.

intolerable when the skin temperature reaches about 37°C (98°F). When γ^* is 0·87, the corresponding equivalent temperatures for a wet surface are 100°C and 109°C. Figure 11.8 shows these limits on a diagram of the type used to present dry heat balances in Chapter 10, but the equivalent temperature θ^* now replaces T as the ordinate. The left-hand side of the diagram shows how θ^* is related to the temperature and vapour pressure of the ambient air. Three lines of constant wet bulb temperatures (10, 20, 30°C) are plotted as a reminder that T' is closely related to θ^* when $\gamma \doteq \gamma^*$.

The right-hand side of the diagram shows gradients of equivalent temperature for a nude and for a clothed subject, both assumed to be sweating freely when **M** is 200 W m^{-2} (light work) and $\mathbf{R}_{ni}$ is 300 W m^{-2} (bright sunshine). The mean value of θ^* for the skin of the nude man is assumed to be 98°C. The equivalent temperature of the air $\theta_a{}^*$ must therefore satisfy the equation

$$\mathbf{H} = \mathbf{R}_{ni} + \mathbf{M} = \frac{\rho c_p(\theta_0{}^* - \theta_a{}^*)}{r_{HR}} \qquad 11.16$$

where $\rho c_p / r_{HR} = 17{\cdot}7$ W m^{-2} °C^{-1}. Thus when $\mathbf{H} = 500$ W m^{-2}, $\theta_0{}^* - \theta_a{}^*$ is 28·5°C, $\theta_a{}^*$ is 70°C and equation 11.16 is represented by the line AB. Reference to the left side of the figure shows the range of air temperature and humidity for which $\theta_a{}^*$ is 76·5°C, e.g. 35°C, 30 mbar. The increase of **H** which would make the same environment seem either 'severe' or 'intolerable' can be derived from the figure by extending the line BA upwards. Conversely if **H** is constant, the effect of increasing $\theta_a{}^*$ by raising T_a or e_a is found by displacing AB upwards without changing its slope.

The effect of clothing can be demonstrated on the same diagram if γ^* is assumed to have the same value between the skin and the surface of the clothing as it has in the free atmosphere and if all the radiation from the environment is assumed to be intercepted at the surface of the clothing. Then the flux of heat from the wet skin (at an equivalent temperature $\theta_0{}^*$) to the surface of the clothing (at $\theta_c{}^*$) is given by

$$\mathbf{M} = \rho c_p(\theta_0{}^* - \theta_c{}^*)/r_c \qquad 11.17$$

The diffusion resistance of the clothing r_c is taken as 2·5 s cm^{-1} (equivalent to about 1 clo) so the factor $(\rho c_p / r_c)$ is 4·8 W m^{-2} °C^{-1}. The line CD drawn with this slope shows that when $\theta_0{}^*$ is 98°C, $\theta_c{}^*$ is 56°C. To represent the gradient of θ^* from the surface of the clothing to the ambient air, the line EF is drawn parallel to AB. The presence of clothing decreases the equilibrium value of $\theta_a{}^*$ from 70°C to 27°C and, because CD is much steeper than AB, a severe or intolerable heat stress would be imposed by a relatively small increase of metabolic rate.

The atmospheric conditions for thermal equilibrium can be represented to a good approximation by a wet bulb temperature of 25°C for the nude subject and of 11°C for the clothed subject. The wet bulb temperature has frequently been used as an index of environmental temperature in human studies. Figure 11.8 confirms that it is a good index for a fully wetted skin provided $r_{HR} \doteq r_V$ but it is inappropriate when the skin temperature is below the limit for rapid sweating.

Temperature enthalpy diagram

The heat balance for an incompletely wetted subject can be derived by appeal to experiments which show that sweat rate increases linearly with metabolic rate **M** and with skin temperature T_0 above a base level T_b.[48] If **P** is the latent heat equivalent of the sweat rate, the relation can be written

$$\mathbf{P} = \mathbf{M} + h(T_0 - T_b) \qquad 11.18$$

where h and T_b vary somewhat from one subject to another. Representative values are $h = 120$ W m^{-2} °C^{-1} and $T_b = 34$°C.

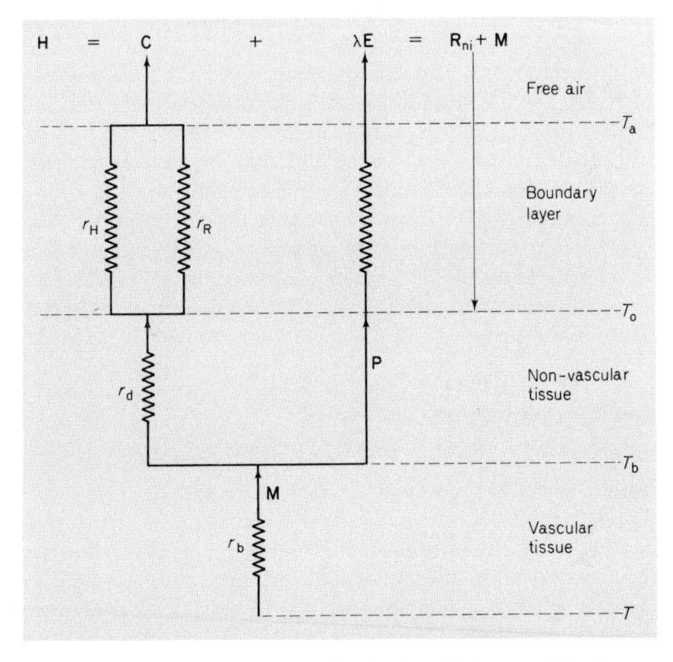

Fig. 11.9 Simple electrical analogue for the heat balance of the human body, distinguishing sensible and latent heat transfer.

Figure 11.9 represents a simple model of the human body based on this equation. The mean temperature of the skin surface is T_0 and T_b is the mean temperature of tissue immediately beneath the non-vascular layer. The thermal resistance of this layer r_d is assumed to be constant whereas the resistance of the rest of the body including the vascular system (r_b) depends mainly on the rate at which blood circulates below the skin.

When the surface of the skin is dry, the evaporation of sweat is assumed to occur in a thin layer of tissue which has a temperature of T_b, so that water vapour must diffuse through the non-vascular layer to reach the skin surface. Then if **P** is the latent heat removed by the evaporation of sweat, the heat balance of the tissue beneath the non-vascular layer is given diagrammatically by Fig. 11.9 and analytically by

$$\mathbf{M} = \frac{\rho c_p}{r_d}(T_b - T_0) + \mathbf{P} \qquad 11.19$$

which is formally similar to equation 11.18 if $h = \rho c_p / r_d$. When $h = 120$ W m^{-2} °C^{-1}, r_d is 0·1 s cm^{-1}. This is the resistance of about 2 mm of skin, consistent with estimates for the thickness of the non-vascular tissue.

Although equation 11.19 is a convenient method of relating measurements of **P**, **M** and T_0, the model has several limitations. It is not valid in principle when the skin is wetted by sweat because the loss of latent heat then occurs at the skin surface and may be smaller than **P**. Even when the skin is dry, the assumption of a constant depth for the sink of latent heat is inconsistent with evidence that the diffusion path for water vapour in the skin increases rapidly with decreasing temperature.

Figure 11.10 shows how the sweat equation can be represented on a TED. It is convenient to replace the fixed temperature T_b by a lower temperature which depends on the metabolic rate and is defined by

$$T_b' = T_b - \mathbf{M} r_d / \rho c_p$$

The sweat equation can now be written

$$\mathbf{P} = \rho c_p (T_0 - T_b') / r_d$$

For example when $\mathbf{M} = 90$ W m^{-2}, $T_b = 34$°C and $\rho c_p / r_d = 120$ W m^{-2} °C^{-1}, T_b' is 33·25°C. This temperature fixes one end of the sweat line (NQ) at 33·25°C on the saturation curve. The slope of the line is chosen so that its distance below the saturation curve (e.g. PQ) is a measure of **P**. More exactly, when r_{HR} is 0·68 s cm^{-1} a difference in equivalent temperature of 1°C represents a heat flux of $(\rho c_p / r_{HR}) \times 1$ or 17·7 W m^{-2} °C^{-1}. So a 1°C increase in skin temperature which releases sweat at a rate corresponding to 120 W m^{-2} can be represented by a change in equivalent temperature of 120/17·7 or 6·8°C. The sweat line $\mathbf{P}_{90}$ is drawn so that its distance below the saturation curve increases by 6·8°C on the equivalent temperature scale when skin temperature increases by 1°C. The line PQ represents the production of 1 litre of sweat per hour by a man with a surface area of 1·8 m^2. The lower pecked line $\mathbf{P}_{360}$ represents the sweat equation when $\mathbf{M} = 360$ W m^{-2} and $T_b' = 31$°C.

When the temperature and vapour of the environment are specified, the sweat rate is determined as follows. The state of the environment is

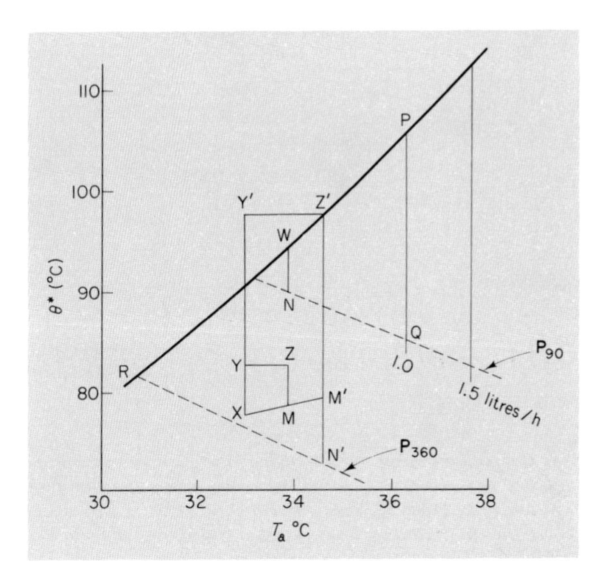

Fig. 11.10 TED for sweating man. The line RZ′ corresponds to saturation vapour pressure and X represents the state of the ambient air. For a relatively small heat load XY (90 W m^{-2}) the latent heat of sweating is given by ZM = WN and the skin is incompletely wetted. For a relatively large heat load XY′ (360 W m^{-2}) the heat equivalent of the sweat rate Z′N′ exceeds the demand for evaporative cooling Z′M′ and sweat drips off the body. The pecked lines $\mathbf{P}_{90}$ and $\mathbf{P}_{360}$ represent the sweat rates for total heat loads of 90 and 360 W m^{-2}.

marked on the TED at X (33°C, 43 mbar in the example). A line XY is drawn corresponding to the total heat load; YZ represents the sensible heat loss and ZM the latent heat loss as in Figs. 11.2 and 11.3. In the first example the heat load is $\mathbf{M}=90$ W m^{-2} so XY $=90/17{\cdot}7=5$°C. The line YZ is projected to the point at which ZM, the evaporation rate at a skin temperature of 33·9°C, is exactly equal to the sweat rate at this temperature shown by WN (0·2 l h^{-1}). If the skin temperature rises above 33·9°C, sweat will be produced more rapidly than it evaporates, but, in principle, a state of thermal equilibrium can be achieved without the skin getting wet.

In the second example, the metabolic rate is increased to 360 W m^{-2} and is represented by XY′. The line Y′Z′ is drawn to intersect the saturation line at Z′ and the rate of sweating Z′N′ (1·2 l h^{-1}) is faster than the rate at which sweat evaporates Z′M′ (0·9 l h^{-1}). In this case thermal equilibrium cannot be achieved unless the whole surface of the body is wet and the air in contact with the surface is saturated. Analytical solutions of this problem have been provided by Hatch[48] and others.

12

The Micrometeorology of Crops

Why do we see the rose bursting out in spring, the corn in scorching summer, the vine at autumn's coaxing, if it is not because, only when the fixed seeds of things have streamed together at their appropriate time, is any created thing uncovered, while the attendant seasons assist the prolific earth to deliver the frail objects into the shores of light in safety?

The application of meteorological principles to the analysis of crop–weather relationships is a branch of agricultural science which has developed rapidly over the last 25 years. Four main problems can be distinguished:

(1) What is the rate at which a crop loses water in a given environment, how does it depend on air, soil and plant factors, and how can the rate be minimized?

(2) What is the rate at which a crop absorbs carbon dioxide in a given environment, how does it depend on air, soil and plant factors and how can the rate be maximized?

(3) What determines the temperature, humidity and wind regime of a crop and what part do these factors play in determining growth and development apart from their effects on transpiration and photosynthesis rates?

(4) How does the microclimate of individual leaves (as distinct from the microclimate of the canopy as a whole) determine the activity of insects and fungi and the development of disorders and diseases which they cause?

Problem (1) has had most attention because the immediate agricultural rewards are greatest. Research on (2) is very active but (3) and (4) are relatively neglected aspects of agricultural meteorology.

Both measurements and models have been used to study the gas exchange of single leaves, of layers of leaves within a canopy and of the canopy as a whole. Figure 12.1 illustrates these different approaches in terms of an electrical analogue. The exchange of any entity between a single leaf and its environment can be estimated (a) when the relevant environmental potential is known (e.g. the air temperature or CO_2 concentration) and (b) when the relevant resistances (e.g. r_s, r_H) can be determined from transfer theory and from a knowledge of the anatomy and physiology of the leaf. In the same way, the bulk exchange of any

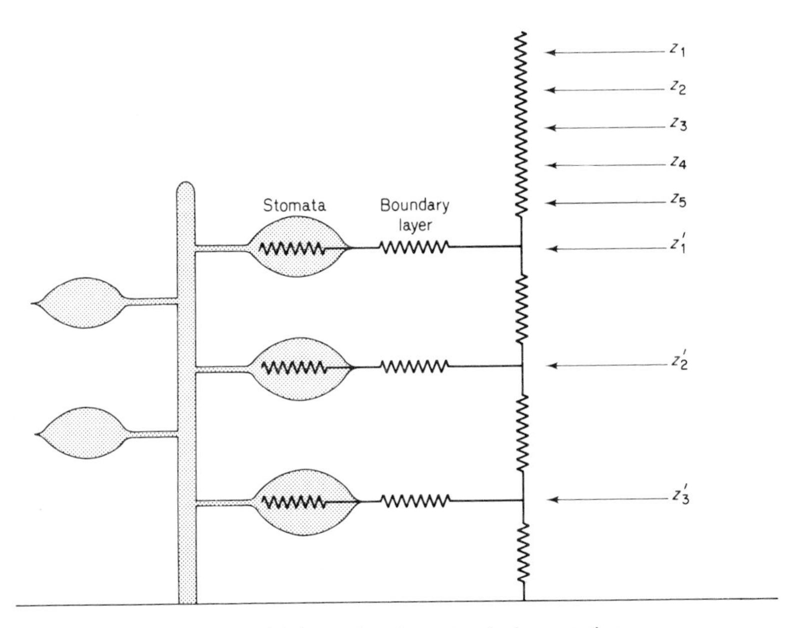

Fig. 12.1 Resistance model for a plant in a stand of vegetation.

entity between a canopy and the air flowing over it can be estimated from the relevant potentials in the air at a single height (i.e. from standard climatological data as in the Penman formula) when the resistances of the system are known (e.g. r_c, r_{aH}). Exchange rates for a canopy can also be determined from the measurement of potentials either *above* a canopy at two or more heights (z_1, z_2, etc.) or *within* a canopy (z_1', z_2', etc.). In both cases, the resistances across which these potentials are measured must be known in order to estimate the corresponding fluxes.

Salient features of measurements and models will now be considered.

MEASUREMENT OF FLUX ABOVE THE CANOPY

Over an extensive uniform stand of level vegetation, there is a layer of air within which fluxes of heat, water vapour and carbon dioxide are constant with height (p. 95). Bulk rates of exchange between the foliage and the air flowing over it can be determined by measuring vertical fluxes in this boundary layer.

There are two principal methods of calculating fluxes above a uniform stand of vegetation from potentials in the boundary layer and they are usually referred to as the 'aerodynamic' and 'Bowen ratio' methods. In both methods, potentials are measured between the top of the crop at height h and the top of the boundary layer at height Z. This height cannot be determined precisely but the measurements on wind profiles discussed on page 96 suggest, as a rule of thumb, that $Z-d$ should not exceed 0·5% of the fetch over a uniform surface with a zero plane displacement of d. In practice, measurements are often taken up to a height where $Z-d$ is 1% or more of the fetch and observations from the top of the profiles are discarded if they are obviously anomalous (e.g. if the potentials are not linearly related to each other).

Fluxes in the boundary layer can also be determined by correlating fluctuations of temperature, humidity and carbon dioxide concentration with fluctuations of vertical velocity but because this technique needs complex sensing and recording equipment it has been applied exclusively to micrometeorological studies of the physical behaviour of the lower atmosphere and flux measurements have usually been restricted to periods of the order of minutes rather than days.

The basis of the **aerodynamic method** was described in Chapter 6. Three sets of profiles are needed: the concentrations of water vapour χ or of carbon dioxide ϕ are measured at a series of heights (shown as $z_1 \ldots z_6$ on Fig. 12.1) using gas analysers or psychrometers and the wind-speed u and temperature T are measured at identical heights with anemometers and a set of thermocouples, thermistors or resistance thermometers. For analysis, the friction velocity u_* is derived from the shapes of the wind and temperature profiles and the fluxes are estimated from

$$\mathbf{E} = -u_*^2 \frac{\partial \chi}{\partial u}$$

$$\mathbf{F} = -u_*^2 \frac{\partial c}{\partial u}$$

In adiabatic conditions, u_* can be estimated from the wind profile without reference to the temperature profile (see p. 91) but in diabatic conditions both profiles are needed (equation 7.19).

The **Bowen ratio** formula for flux measurement is derived from the energy balance of the underlying surface

$$\mathbf{R}_n - \mathbf{G} = \mathbf{C} + \boldsymbol{\lambda}\mathbf{E}$$

which can be rewritten in the form

$$\lambda \mathrm{E} = \frac{(\mathbf{R}_n - \mathbf{G})}{1+\beta} \qquad 12.1$$

where β is the Bowen ratio $\mathbf{C}/\boldsymbol{\lambda}\mathbf{E}$. Measurements of the net radiation ($\mathbf{R}_n$) and soil heat flux ($\mathbf{G}$) are needed to establish ($\mathbf{R}_n - \mathbf{G}$) and β is found from measurements of temperature and vapour pressure at a series of heights within the boundary layer of the surface. Assuming that the transfer coefficients of heat and water vapour are equal, it can be shown that

$$\beta = \mathbf{C}/\boldsymbol{\lambda}\mathbf{E} = \gamma\, \partial T/\partial e \qquad 12.2$$

and $\partial T/\partial e$ is found by plotting the temperature at each height against vapour pressure at the same height.

The Bowen ratio method can be generalized by writing the heat balance equation as

$$\begin{aligned} \mathbf{R}_n - \mathbf{G} &= -K\rho c_p(\partial T/\partial z) - K(\partial \chi/\partial z) \\ &= -K\rho c_p(\partial \theta/\partial z) \end{aligned} \qquad 12.3$$

where K is a turbulent transfer coefficient assumed identical for heat and water vapour and θ is the equivalent temperature. Similar expressions relate the sensible heat flux $\mathbf{C}$ to the temperature gradient $\partial T/\partial z$, the latent heat flux $\boldsymbol{\lambda}\mathbf{E}$ to $1/\gamma(\partial e/\partial z)$ and the carbon dioxide flux $\mathbf{F}$ to $\partial \theta/\partial z$. Combining these expressions in turn with equation 12.3, it can readily be shown that

$$\mathbf{C} = (\mathbf{R}_n - \mathbf{G})(\partial T/\partial \theta) \qquad 12.4a$$

$$\boldsymbol{\lambda}\mathbf{E} = (\mathbf{R}_n - \mathbf{G})(\partial e/\partial \theta)/\gamma \qquad 12.4b$$

$$\mathbf{F} = (\mathbf{R}_n - \mathbf{G})(\partial \phi/\partial \theta)/\rho c_p \qquad 12.4c$$

The Bowen ratio equation 12.1 is derived from equation 12.3 by writing $\theta = T + (e/\gamma)$.

The aerodynamic and Bowen ratio methods of flux determinations are usually applied to potentials which have been averaged for periods of a half to one hour. Fluctuations in the potentials, especially on a day of intermittent cloud, often preclude the estimation of mean fluxes for shorter periods. On the other hand, diurnal changes make time averaging undesirable for periods of more than two hours, particularly near sunrise and sunset.

Examples of the calculation of fluxes over a field of barley are given for one hour of a sunny day in Table 12.1 and Figs. 12.2 and 12.3.

Table 12.1 Calculations of fluxes from gradients and equations 12.4 a–c

Measured fluxes ($W\ m^{-2}$)	Measured gradients (Fig. 12.3)
$R_n = 581$	$\frac{\partial T}{\partial \theta} = 0{\cdot}70$
$G = 45$	$\frac{1}{\gamma}\frac{\partial e}{\partial \theta} = 0{\cdot}30$
$R_n - G = 536$	$\frac{\partial \phi}{\partial \theta} = 2{\cdot}2\ mg\ m^{-3}\ {}^{\circ}C^{-1}$

Calculated fluxes

$$C = (R_n - G)\frac{\partial T}{\partial \theta} = 536 \times 0{\cdot}7 = 375\ W\ m^{-2}$$

$$\lambda E = (R_n - G)\frac{1}{\gamma}\frac{\partial e}{\partial \theta} = 536 \times 0{\cdot}3 = 161\ W\ m^{-2}$$

$$F = \frac{1}{\rho c_p}(R_n - G)\frac{\partial \phi}{\partial \theta} = \frac{536 \times 2{\cdot}2}{1{\cdot}2 \times 10^3} = 0{\cdot}98\ mg\ m^{-2}\ s^{-1}$$

$$= 3{\cdot}5\ g\ m^{-2}\ h^{-1}$$

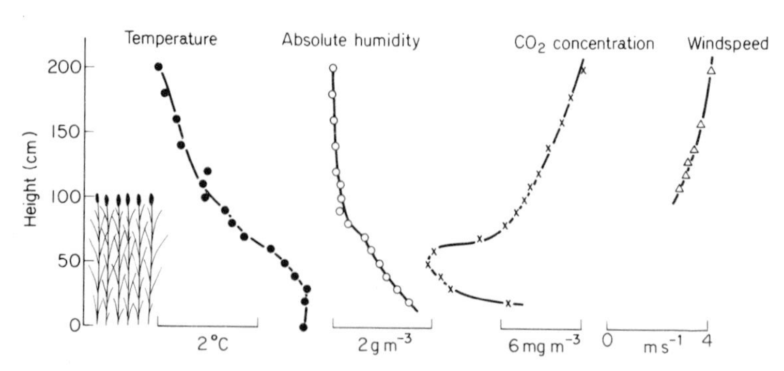

Fig. 12.2 Profiles in a field of barley at Sutton Bonington, 0900–1000 25 June 1970. The stand height was approximately 95 cm.

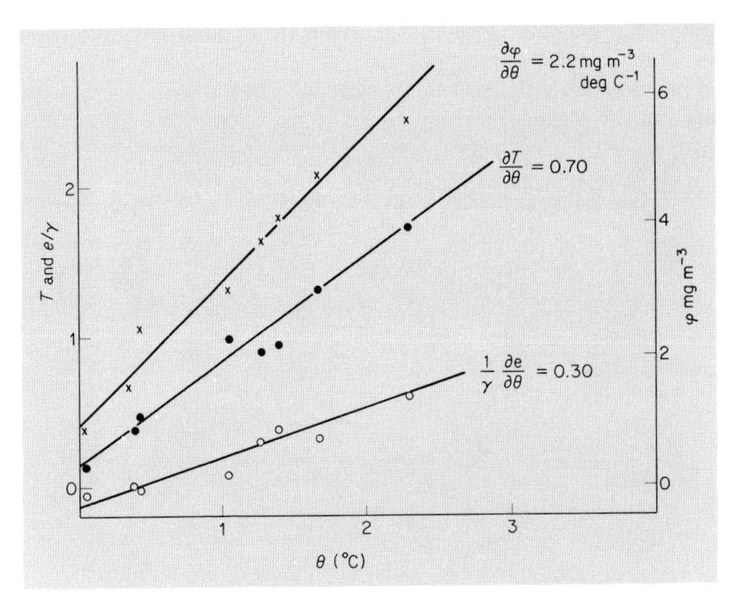

Fig. 12.3 The relation between T, e, ϕ and θ for the profiles in Fig. 12.2 using measurements above the canopy only.

INTERPRETATION OF FLUX MEASUREMENTS ABOVE THE CANOPY

Water vapour

The aerodynamic method of measuring water vapour fluxes, first derived more than thirty years ago, has become a standard method of estimating evaporation from lakes and reservoirs. It has seldom been used to measure the loss of water from vegetation in agricultural or ecological studies and there are several reasons why the method has not commended itself to physical ecologists. In the first place, constant vigilance is needed to maintain instruments in first class order for periods comparable with the growing season of a crop and the records that accumulate are tedious to analyse unless they are available in a form that can be digested by a computer. In the second place, a knowledge of the rate of transpiration from a given crop in a given environment has little prognostic value until it is associated with some parameter or group of parameters describing the physiological control of water loss. It is possible to derive a parameter which plays the same part in equations for the water vapour exchange of a canopy as the stomatal resistance r_s plays in the similar equations for a single leaf. This parameter will be given the symbol r_c where the subscript denotes canopy, crop or cover.

It has been shown that the sensible heat loss from a surface can be written in the form

$$\mathbf{C} = -\rho c_p u_*^2(\partial T/\partial u)$$

where T is a linear function of u. As a special case, the gradient $\partial T/\partial u$

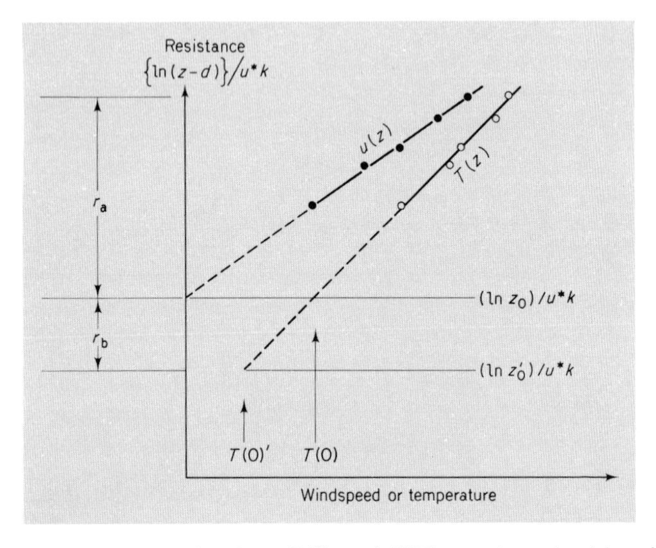

Fig. 12.4 Diagram showing how $T(0)$ and $T(0')$ are determined by plotting temperature against $[\ln(z-d)]/u_*k$. $T(0)$ is the value of temperature extrapolated to $z-d=z_0$ and $T(0')$ is the value extrapolated to $z-d=z_{0'}$. The significance of r_a and r_b is shown on the left-hand axis and is discussed on p. 198.

can be written $[T(z)-T(0)]/[u(z)-0]$ where $T(0)$, obtained by the extrapolation shown in Fig. 12.4, is the air temperature at the height where $u=0$, i.e. $z=d+z_0$. The above equation can therefore be written as

$$\mathbf{C} = \rho c_p u_*^2[T(z)-T(0)]/u(z)$$

$$= \rho c_p[T(z)-T(0)]/r_{aH} \qquad 12.5$$

where $r_{aH}=u(z)/u_*^2$ can be regarded as an aerodynamic resistance between a fictitious surface at the height $d+z_0$, and the height z. Similarly, it can be shown that

$$\lambda\mathbf{E} = \frac{\rho c_p}{\gamma}\frac{[e(z)-e(0)]}{r_{aV}} \qquad 12.6$$

where $e(0)$ is the value of the vapour pressure extrapolated to $u=0$ and $r_{aV}=r_{aH}=u(z)/u_*^2$. The diffusion of water vapour between the intercellular spaces of leaves and the atmosphere at height z can now be described formally by the equation

$$\lambda \mathbf{E} = \frac{\rho c_p}{\gamma} \frac{[e(z)-e_s T(0)]}{r_{aV}+r_c} \qquad 12.7$$

This relation defines the canopy resistance r_c and is identical to the corresponding equation for an amphistomatous leaf with r_c replacing the stomatal resistance r_s[86] (p. 180) and r_{aV} replacing r_V. Introducing the energy balance equation, writing $r_{aH}=r_{aV}=r_a$ and eliminating $T(0)$ gives

$$\lambda \mathbf{E} = \frac{\Delta(\mathbf{R}_n-\mathbf{G})+\rho c_p[e_s(T(z))-e(z)]/r_a}{\Delta+\gamma^*} \qquad 12.8$$

where $\gamma^*=\gamma(r_a+r_c)/r_a$. Values of r_c for a given stand can therefore be derived either directly from profiles of temperature, humidity and wind using equations 12.6 and 12.7 or indirectly from the Penman equation (12.8) when the relevant climatological parameters are known and $\lambda \mathbf{E}$ is measured or estimated independently.

Two objections can be raised to this apparently straightforward method of separating the aerodynamic and physiological resistances of a crop canopy. In the first place, the values of r_c derived from measurements are not unique unless the sources (or sinks) of sensible and latent heat have the same spatial distribution. In a closed canopy, fluxes of both heat and water vapour are dictated by the absorption of radiation by the foliage and, provided the stomatal resistance of leaves does not change violently with depth in the part of the canopy where most of the radiation is absorbed, the distribution of heat and vapour sources will usually be similar but will seldom be identical. Conversely, anomalous values of r_c are likely to be obtained in a crop with little foliage if evaporation from bare soil beneath the leaves makes a substantial contribution to the total flux of water vapour.

In the second place, the analysis cannot yield values of r_c which are strictly independent of r_a unless the apparent sources of heat and water vapour, as determined from the relevant profiles, are at the same level $d+z_0$ as the apparent sink for momentum. This is a more serious restriction. Form drag, rather than skin friction (p. 81), is often the dominant mechanism for the absorption of momentum by vegetation so that the resistance to the exchange of momentum between a leaf and the surrounding air is smaller than the corresponding resistances to the exchange of heat and vapour which depends on molecular diffusion alone. It follows that the apparent sources of heat and water vapour will, in general, be found at a lower level in the canopy than the apparent sink of momentum, say at $z = d + z_0'$ rather than at $z = d + z_0$ where z_0' is smaller than z_0. The resistance between a height z above the ground and the apparent source of heat and vapour can be found from equation 6.18 as

$$r_a' = \frac{\ln[(z-d)/z_0']}{u_* k} = \frac{\ln[(z-d)/z_0]}{u_* k} + \frac{\ln[z_0/z_0']}{u_* k}$$

$$= \quad r_a \quad + \quad r_b \qquad 12.9$$

where r_a is the resistance for momentum transfer and r_b is the additional boundary layer resistance assumed to be the same for heat and vapour transfer. The resistance r_b is $[\ln(z_0/z_0')]/u_* k$ and $u_* r_b$ is identical to the parameter B^{-1} which a number of workers have used to analyse processes of exchange at rough surfaces. The implications of equation 12.9 are shown in Fig. 12.4.

The size of r_b for real and for model vegetation has been estimated by Chamberlain[22] and Thom[139] and the results of their work can be summarized as follows:

(i) For a given value of u_*, z_0/z_0' and r_b are almost constant over a wide range of surface roughnesses. For example, a set of measurements of evaporation from an artificial grass surface gave $z_0 = 1$ cm, $u_* = 25$ s cm^{-1}, $B^{-1} = 4{\cdot}5$. At the same value of u_*, achieved at a higher windspeed over towelling with $z_0 = 0{\cdot}045$ cm, B^{-1} was 4·7. The corresponding resistances r_b are 0·18 and 0·19 s cm^{-1}.

(ii) For a given value of z_0, z_0/z_0' increases with windspeed and therefore with u_*. For a fourfold increase of u_* from 25 to 100 s cm^{-1}, $u_* r_b$ increased by a factor of 1·3 for the grass and 1·7 for the towelling. For a bean crop, Thom found that $u_* r_b$ was proportional to $u_*^{1/3}$, implying that $r_b \propto u_*^{-2/3}$.

(iii) The value of z_0' and hence of r_b is expected to depend on the molecular diffusivity of the property being transferred. On the assumption that $r_b \propto (\text{diffusivity})^n$, values of n determined experimentally range from about $-0{\cdot}8$ to about $-0{\cdot}3$. For a stand of beans, n appears to be about $-0{\cdot}66$ implying that r_b for heat may be 10% greater than r_b for water vapour. This difference is too small and too uncertain to be considered in the analysis of field data.

The surface resistance of vegetation has sometimes been determined from relations such as

$$\lambda E = \frac{\rho c_p[e_s(T_0) - e]}{\gamma(r_a + r_c)} \qquad 12.10$$

where $r_a = u/u_*^2$, r_c is the 'apparent' surface resistance and $T(0)$ the 'apparent' surface temperature. More rigorous analysis gives

$$\lambda\mathbf{E} = \frac{\rho c_p[e_s(T_0') - e]}{\gamma(r_a' + r_c')}$$

where r_c' is the 'true' resistance allowing for the existence of the additional boundary layer resistance r_b, and T_0' is the 'true' surface temperature. By manipulating these equations and using the relation $\mathbf{C} = \rho c_p(T_0' - T_0)/r_b$, it can be shown that the error in calculating r_c without allowing for r_b is

$$\delta r_c = r_c' - r_c = r_b\left(\frac{\Delta}{\gamma}\cdot\frac{\mathbf{C}}{\lambda\mathbf{E}} - 1\right) \quad 12.11$$

This error is zero when the Bowen ratio $\mathbf{C}/\lambda\mathbf{E}$ is equal to γ/Δ. For a well watered crop growing in a temperate climate, the average value of $\mathbf{C}/\lambda\mathbf{E}$ is usually about 0·1 and with $(\Delta/\gamma) = 2{\cdot}0$, $\delta r_c = -0{\cdot}8 r_b$. The absolute magnitude of this error is less important than the fact that it will change in size and sign during the day as the Bowen ratio changes. For example if $\mathbf{C}/\lambda\mathbf{E}$ decreases from $+0{\cdot}3$ in the early morning to $-0{\cdot}3$ in the late afternoon and r_b is 0·2 s cm^{-1}, the value of r_c will change during the day from $r_c' + 0{\cdot}08$ to $r_c' + 0{\cdot}32$ s cm^{-1}.

Minimum values of r_c reported in the literature[86] are about 0·4 s cm^{-1} and the corresponding value of the corrected resistance r_c' should probably be about 0·2 to 0·3 s cm^{-1}. Measurements on barley[93] (*Hordeum vulgare*) and beans[11] (*Phaseolus vulgaris*) suggest that in a canopy, where the leaves are not densely packed, the main diffusion resistance in the canopy is within the stomata, rather than in the air between plants. In this case, r_c' is formed by a whole set of leaf resistances in parallel so that if leaves with a stomatal resistance of r_s (on each surface) form a canopy with a leaf area index of L, r_c' will be approximately equal to $r_s/2L$. For a stand of barley, the value of r_c' was close to $3/2L$ s cm^{-1} from the middle of May to the end of June.

It has also been shown that r_c increases during the day at a rate that is consistent with porometer measurements of leaf diffusion resistance, that it decreases with increasing irradiance in accord with porometer measurements, that it increases systematically with soil water deficit and that it is almost independent of windspeed.[93, 132] The experimental evidence therefore supports the use of r_c, or better of r_c', as an index of the physiological control of water loss by a crop canopy. No more appropriate index has yet been devised despite attacks which are based on armchair speculation divorced from experience in the field.[106]

Carbon dioxide

The flux of CO_2 in the air above a crop canopy is a measure of the net exchange of CO_2 between the soil–plant system and the atmosphere. Figure 12.5 shows how the components of this exchange are likely to alter over a period of 24 hours.[82] The line zasbz′ represents the flux in the air above a crop, directed upwards during the night when there is a net loss of CO_2 from the system and downwards during the day when there is a net gain. The axis OO′ represents zero flux. Respiration from three sources contributes to the flux of CO_2 at night: plant tops, plant roots and soil organisms. Total respiration between midnight and sunrise is represented by za. At sunrise (a) the photosynthetic system begins to assimilate part of the respired CO_2 and the upward flux decreases to zero when solar irradiance reaches the light compensation point for the stand, usually about 1 to 2 hours after sunrise over actively growing vegetation. After the irradiance exceeds the compensation point, there is a downward flux of CO_2 representing the atmospheric contribution to photosynthesis. Shortly before sunset (b) the compensation point is reached again and after sunset the rate of respiration is shown by bz′.

Components of the CO_2 balance during the day are given by segments of the line sw:

st = uptake of CO_2 from atmosphere
tw = uptake of CO_2 from plant and soil respiration
sw = gross uptake of CO_2
uw = plant respiration
su = net photosynthesis

Only one of these quantities, st, can be readily measured. The respiration during the day is not known but can be estimated from the average flux

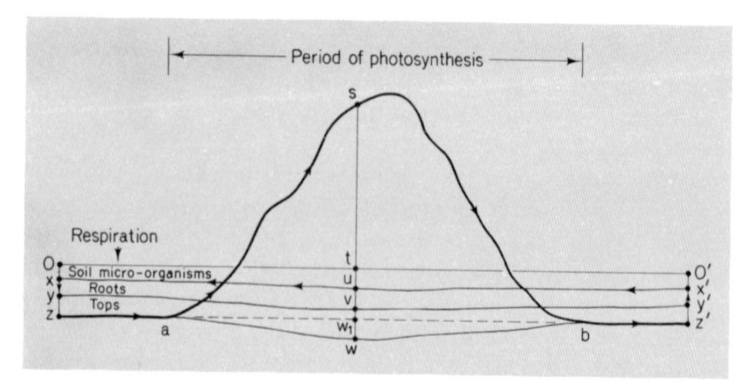

Fig. 12.5 The diurnal change of CO_2 flux above a stand of vegetation, shown by the bold line zasbz′. For the significance of other components, see text.

at night obtained by drawing a straight line through za and bz′ intersecting sw at w_1. The segment ww_1 represents the increase in total respiration as a result of the higher soil and air temperature during the day. The proportion of total respiration attributable to soil organisms is very difficult to establish experimentally because the presence of plant roots stimulates microbial activity in the rhizosphere.[92] If β is the ratio of plant respiration to the total respiration of the system, the instantaneous rate of net photosynthesis is $(sw_1) - \beta(tw_1)$. During the life of a crop, the value of β will increase from zero at germination to a maximum which will usually lie between 0·5 and 0·9 when the crop is mature.

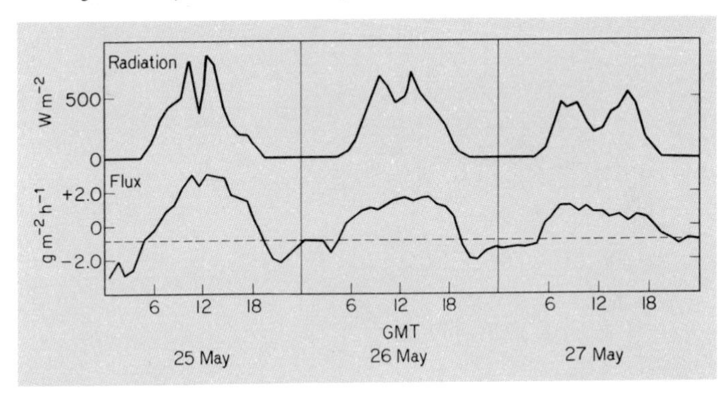

Fig. 12.6 The diurnal variation of CO_2 flux over a stand of grass and total solar radiation. The pecked line is the estimated respiration rate for calculating gross photosynthesis.

The integrated rates of photosynthesis for the 24 hour period are

gross photosynthesis: area asbwa ≏ $asbw_1a$
plant respiration: area xux′z′wz
net photosynthesis: area zsz′x′ux (which would be measured by a planimeter following the arrows on the Fig. 12.5)

In practice, the net photosynthesis would need to be found from the area zszO′Oz plus the nocturnal respiration from soil organisms which is $(1-\beta)$ times the total nocturnal respiration.

As a warning that Fig. 12.5 is very idealized, Fig. 12.6 shows the real diurnal change of flux over grass on three consecutive days in May, 1961.[82] The estimates of flux were uncorrected for atmospheric stability so the nocturnal estimates of respiration are too large except during relatively windy hours at the beginning and end of the period. The pecked line (corresponding to zz′) shows the best estimate that could be made of respiration. When β is assumed to be 1·0 and hourly gross photosynthesis is plotted against solar radiation, the points form two distinct light response curves for two of the days (Fig. 12.7). The analysis illustrates the kind of information which atmospheric measurements of flux can provide without interfering in any way with the microclimate or growth of the crop.

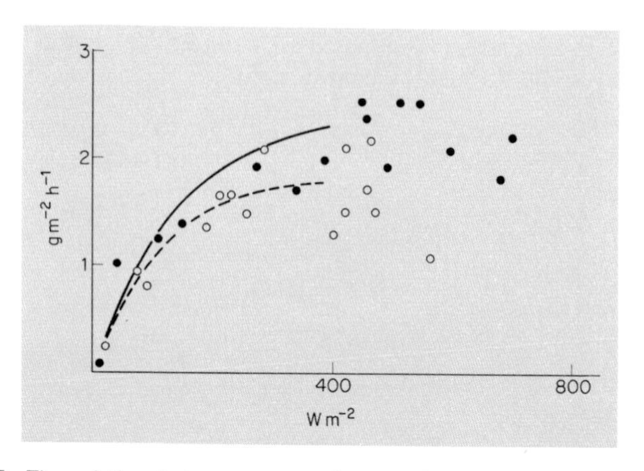

Fig. 12.7 The relation between gross photosynthesis of grass and total solar radiation. Filled circles 26.5.61, open circles 27.5.61. The pecked line was derived from laboratory measurements on a single layer of leaves and the continuous line was estimated for two layers of leaves.

MICROCLIMATE OF CROPS

The aerodynamic and Bowen ratio methods of flux estimation use measurements of potentials in the boundary layer *above* the crop canopy. No measurements are made *within* the canopy although it is possible to derive bulk parameters such as z_0, d and r_c related to the geometry and physiology of the foliage. Information about the response of a crop to its environment can also be derived from the nature of its microclimate—the regime of radiation, temperature, humidity, carbon dioxide and wind between the top of the canopy and the soil surface. Many measurements of microclimate have been published and Table 12.2 summarizes recent literature for different types of vegetation.

Salient features of microclimate are illustrated in Fig. 12.8 where a series of profiles represents the conditions within a cereal crop growing to a height of $h = 1$ m with most of its green foliage between $h/2$ and h.

Table 12.2 Literature on crop micrometeorology

Crop	Solar radiation	Net radiation	Air temperature and heat flux	Soil temperature and heat flux	Water vapour	Carbon dioxide	Wind and momentum
Barley (*Hordeum vulgare*)	Pi (69)	Pi (69)	Pai (69)		Pai (69)	Pai (69)	PaFa (69)
Bulrush millet (*Pennisetum typhoides*)		Pi (7)	Pi (7)		PFi (7)		
Cotton (*Gossypium barbadense*)	Fa (116, 128)	Fa (117, 128)		F (128)			Fa (128)
Lucerne (*Medicago sativa*)	Fa (119)		PFa (119)		PFa (119)	PFa (119)	Pa (119)
Maize (*Zea mays*)	Pi (145)	Pi (17)	PFi (17)		PFi (17)	Pia (144) Fi (54) PFa (63) PFi (65)	PFai (143)
Orange (*Citrus sinensis*)	Fa (57)	PFia (57)	Pi (57)		Pi (57)		Pia (57)
Potato (*Solanum tuberosum*)			Pia (70, 76, 149)	PF (149)	Pia (70, 149)		Pia (70)
Sugar beet (*Beta vulgaris*)			Pia (75)		Pia (75)	Pia (134) Fa (90)	
Sunflower (*Helianthus annuus*)	Fa (37)		PFa (37)		PFa (37)	PFa (37)	Pa (37)
Wheat (*Triticum aestivum*)		Fa (31)	Pia (75, 103)	PF (75, 149)	Pia (103) Fa (31)	PFa (31)	Pia (103)

Key: F—Flux; P—Potential; a—above canopy; i—inside canopy

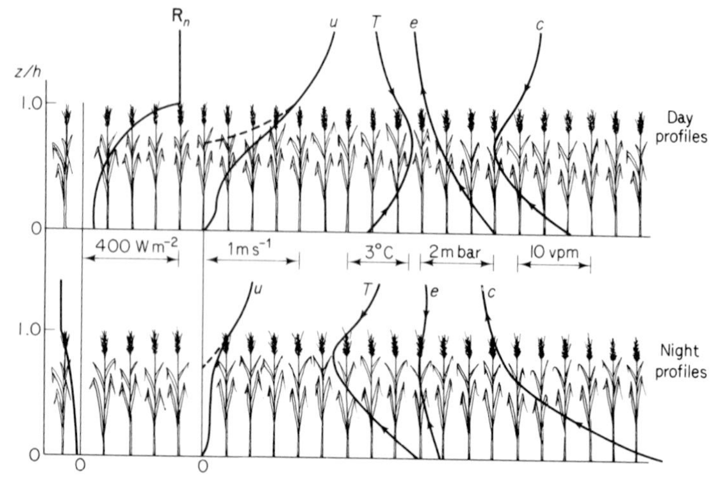

Fig. 12.8 Idealized profiles of net radiation ($\mathbf{R}_n$), windspeed (u), air temperature (T), vapour pressure (e) and CO_2 concentration (c) in a field crop growing to a height h plotted as a function of z/h. The pecked wind profiles represent an extrapolation of the logarithmic relation between u and $(z-d)$ above the canopy.

Radiation

Solar radiation is absorbed rapidly by foliage and the relation between irradiance and cumulative leaf area is almost exponential[89] (p. 52). Because leaves absorb visible light strongly but transmit long wave radiation between 0·7 and 1 μm, visible radiation is attenuated more rapidly than solar radiation and is often reduced to between 5 and 10% of the external irradiance beneath a mature crop. The leaf area index needed to produce 95% attenuation of visible radiation ranges from about 3 for stands with horizontal leaves to 10 for stands with more vertical leaves. In forests where branches and trunks intercept a significant amount of light, the average irradiance of visible light on the floor can be as small as 1 or 2% of the external flux.

In the upper part of a canopy during the day, the shape of the net radiation profile is usually very similar to the shape of the solar radiation profile but the profiles tend to differ close to the soil unless the soil temperature is very similar to the temperature of the lowest foliage. At night, the net radiation profile is reversed and is sensitive to changes of temperature gradient within the canopy.

Wind

Above the canopy, windspeed is a logarithmic function of height above the zero plane (p. 88). In the top part of the canopy, but below the top of the crop, the real windspeed (Fig. 12.8, continuous line) exceeds the value predicted by the logarithmic relation (pecked line) and may even increase slightly near the soil surface. This effect has been observed with several different types of anemometers both in crops and in forests and is probably a consequence of horizontal momentum transfer beneath the main layer of foliage rather than vertical transfer through it. Equations for the shape of the wind profile in canopies are discussed on p. 209.

Temperature

The slope of temperature profiles is a direct consequence of the heat balance of the system and indicates the direction of heat flow at each level (shown by arrows in Fig. 12.8). When a crop is freely supplied with water so that the stomatal resistance of the upper leaves is small (say 1–3 s cm^{-1}), there is often a temperature maximum near the top of the foliage during the morning (associated with an upward flux of sensible heat from surface to air) and a less marked minimum at about the same level in the late afternoon (downward flux of heat). When the water supply is restricted and r_c is large, the maximum persists later in the day and is lower in the canopy, sometimes reaching the soil surface. On a clear night there is a minimum near the top of the canopy which tends to disappear if the sky becomes overcast.

Vapour pressure

The shape of vapour pressure profiles, like the temperature profiles, is a consequence of the heat and water balance of the system and reveals the direction of water vapour flux. During the day the vapour pressure usually decreases from the soil surface upwards. There is a much smaller decrease at night sometimes associated with a minimum vapour pressure near the top of the canopy when dew is forming.

Carbon dioxide

In daylight there is a minimum concentration, often between 270 and 290 vpm, near the top of the canopy. At night there is a maximum, often exceeding 400 vpm near the soil surface.

Analysis of microclimate

The shapes of profiles within a canopy can be used to determine the distribution of sources and sinks of heat, water vapour and carbon dioxide. The *location* of sources and sinks can often be established by inspection of the profiles but the *strength* of sources and sinks cannot be calculated without knowing how the turbulent transfer coefficient K changes with height within the canopy.

In this context, a 'source' of heat (or water vapour or CO_2) in a canopy is equivalent to a layer where the vertical flux diverges, i.e. the flux leaving the layer exceeds the flux entering the layer. The transfer equations are formally identical to those which describe heat transfer in the soil, (p. 128). If a positive value of $\mathbf{C}$ represents an upward flux of heat, then a layer in the canopy where $\partial\mathbf{C}/\partial z$ is positive will be a source of heat and a layer where $\partial\mathbf{C}/\partial z$ is negative will be a sink. Note that a positive value of $\partial\mathbf{C}/\partial z$ can be achieved (a) if an upward flux increases with height or (b) if a downward flux decreases with height. If ρc_p is the volumetric specific heat of air and K, a function of height, is the transfer coefficient in the canopy, the source strength can be written as

$$\frac{\partial\mathbf{C}}{\partial z} = -\frac{\partial}{\partial z}\left(\rho c_p K \frac{\partial T}{\partial z}\right) = -\left\{\rho c_p \frac{\partial K}{\partial z}\frac{\partial T}{\partial z} + K\frac{\partial T}{\partial z}\right\} \qquad 12.12$$

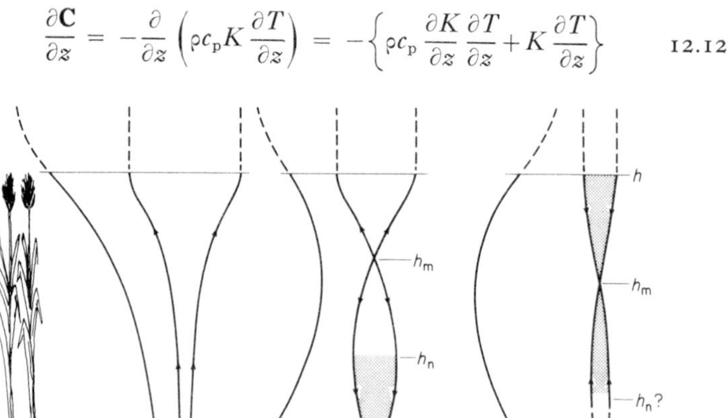

Fig. 12.9 Hypothetical distribution of a potential and flux in a cereal crop. In each of the three sections a, b and c, the left-hand curve represents the potential and the separation of the right-hand pair of curves is proportional to the flux; h is crop height, h_m is height at which flux is zero and h_n is height where flux divergence is zero. The stippled areas represent heights at which the flux of an entity is being absorbed by the vegetation.

The position of sources can often be determined without evaluating the terms in this equation if K is assumed to increase between the soil surface and the top of the canopy, i.e. $\partial K/\partial z$ is positive throughout the canopy. A layer in which the two terms in curly brackets are both negative must be a source of heat because $\partial\mathbf{C}/\partial z$ is positive, and conversely a layer in which both terms are positive must be a sink. When the terms have opposite signs, they must be evaluated to determine the sign of $(\partial\mathbf{C}/\partial z)$.

The interpretation of the profile of any potential in terms of the location of sources and sinks is shown graphically in Fig. 12.9. Beside the three profiles (a), (b) and (c) there are three pairs of lines which are mirror images. The arrows on these lines show the direction of the fluxes depending on the way in which the potential changes with height, and the separation of the lines is proportional to the size of the flux.

In (a), the flux increases steadily with height to the top of the canopy (cf. the profile for water vapour during the day or for CO_2 at night, Fig. 12.8). In (b), the flux is zero at a height h_m and a layer of foliage at this level acts as a source from which heat or mass diffuses upwards and downwards (cf. the profile for temperature during the day). The profile below the level h_n acts as a sink. In profile (c) a layer of foliage at a height h_m acts as sink (cf. the profile for CO_2 during the day). A source *may* exist below h_n but equation 12.12 is indeterminate in this case. The interpretation of profiles in this way is valid only for horizontally homogeneous stands of vegetation in which the processes of exchange are dominated by the convergence and divergence of *vertical* fluxes. In small plots, microclimates may be determined by the rate at which horizontal fluxes change with distance from the edge of the plot as well as the rate at which vertical fluxes change with height.

MEASUREMENT OF CANOPY FLUXES

The calculation of source and sink strengths in a canopy is bedevilled by two awkward features of equation 12.12. First, it is necessary to know the transfer coefficient K and its height variation $\partial K/\partial z$. Second, it is necessary to determine the profile shapes very precisely so that the second differential of the relevant potential can be estimated with enough precision. Several methods have been proposed for determining K as a function of height and two will be considered here.

Momentum balance

Within any layer of canopy where the measured horizontal windspeed is u, the drag force exerted on a leaf with area A is $F=\frac{1}{2}c_d\rho u^2A$, where c_d is a drag coefficient (p. 82). If the leaf area per unit depth of canopy is l, the drag in an infinitesimal layer dz can be written $df=\frac{1}{2}c_d\rho u^2l\,dz$, so that the total drag is

$$\int_0^h df = \tfrac{1}{2}\rho\int_0^h c_d l u^2\,dz \qquad 12.13$$

The total drag can also be measured from the wind profile above the canopy: it is $\tau=\rho u_*^2$. The equation

$$\rho u_*^2 = \tfrac{1}{2}\rho\int_0^h c_d l u^2\,dz \qquad 12.14$$

must therefore be satisfied for all values of u and u_*. An approximate solution of the equation can be obtained by assuming that c_d is constant. (This condition is met only when the force exerted by skin friction is much smaller than form drag and no streamlining occurs.) The constant value of c_d can then be determined as

$$c_d = u_*^2 \Big/ \left(\tfrac{1}{2} \int_0^h lu^2 \, dz \right)$$

This method has the practical disadvantage of needing detailed measurements of the distribution of leaf area with height but a much more serious objection arises from the tacit assumption that the leaves are exposed to the windspeed u that would be recorded by an anemometer at the same height. If the leaves tended to shelter each other, the effective windspeed could be substantially smaller than u and an error of this kind has important implications for the behaviour of the drag coefficient. The windspeed recorded in crop canopies is often about 0·3 to 0·6 m s^{-1} and in this range the assumption that c_d is independent of windspeed is approximately true. If the effect of mutual sheltering makes the effective windspeed say 0·1 to 0·2 m s^{-1}, the drag coefficient will be larger and will change much more rapidly with windspeed because of the importance of skin friction in this range (Fig. 6.4).

The size of the shelter effect in a stand of beans (*Vicia faba*) was demonstrated by Thom[138] from wind tunnel measurements of the drag on leaves and from estimates of the contribution to the drag from pods and petioles. The drag force per unit area on individual leaves was assumed to be $\frac{1}{2}\rho c_d u^2/p$ where p is a shelter factor and, by equating the integrated drag to ρu_*^2, p was found to be 3·5. The drag coefficient was not independent of windspeed but was approximately proportional to $u^{-1/2}$, implying that skin friction was more important than form drag.

Whatever method is used to establish an appropriate value of c_d for foliage, the transfer coefficient K_M for momentum at each height in the canopy can be determined from equation 12.14 when c_d, $u(z)$ and $l(z)$ are known. When c_d is assumed to be constant, estimated values of K decrease almost exponentially with height and several workers have fitted their results to the equation

$$K = K_h \, e^{-n(1 - z/h)} \qquad 12.15$$

where K_h is the value of K at the top of the canopy $z = h$. Values of n for wheat, rice, clover and corn lie between 2 and 3. Cowan[27] derived a more complex exponential relation between K and z/h starting with the assumption that profiles of u and of K had the same shape in the canopy.

When proper account is taken of the shelter effect and of the way in which c_d increases with depth in the canopy as a result of decreasing windspeed, estimated values of K are larger and change more slowly with height than the values predicted from equation 12.15. In fact, Thom found that K was almost constant in a bean crop (*Vicia faba*) between $h/3$ and h (Fig. 12.10). His analysis was based on a drag coefficient c_d proportional to $u^{-1/2}$ (the relation for skin friction). Nevertheless, with

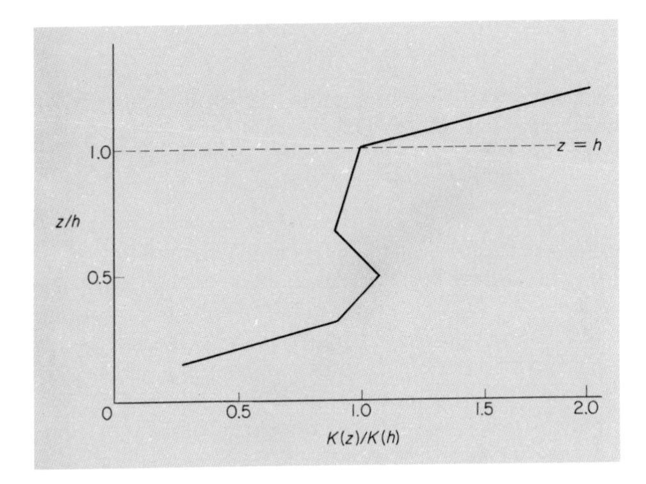

Fig. 12.10 Change of transfer coefficient $K(z)$ with height in a field of beans (*Vicia faba*) calculated from drag coefficients and wind profiles measured by I. F. Long at Rothamsted, 7 July 1966 (from Thom[138]).

the basic assumption that K is independent of z, and the algebraic simplifications that l and c_d are constant, it can be shown that the wind profile is given by

$$u = u(h)\{1+\alpha(1-z/h)\}^{-2} \qquad 12.16$$

For the beans, equation 12.16 provides a good fit to measure wind profiles when the coefficient α has the value 1.3. Values for other types of stand have not yet been evaluated.

The shape of the wind profile derived theoretically is not sensitive to assumptions about the behaviour of the drag coefficient or the size of the shelter factor. Almost identical profiles can be obtained taking c_d as constant and putting $p=1$ or assuming $c_d \propto u^{-1/2}$ and $p=3{\cdot}5$. In contrast, the profiles of K derived from the two premises have a completely different form. As a corollary, it is extremely difficult to deduce the shape of the K profile accurately from the shape of the wind profile but relatively easy to simulate a realistic wind profile from assumptions about the increase of K with height.

Energy balance

Many of the objections to determining $K(z)$ from the momentum balance can be overcome by applying the energy balance to a layer of crop between a height z and the soil surface.[157] The net gain of heat by this layer is $\mathbf{R}_n(z)-\mathbf{G}$ where $\mathbf{R}_n$ is the (downward) flux of net radiation at z and the net loss by (upward) turbulent transfer is

$$\mathbf{C}+\lambda\mathbf{E} = -\rho c_p K(\partial\theta/\partial z)$$

where θ is the equivalent temperature. When the layer is in thermal equilibrium so that $\mathbf{R}_n-\mathbf{G}=\mathbf{C}+\lambda\mathbf{E}$ the value of $K(z)$ can be found from

$$K = (\mathbf{R}_n-\mathbf{G})/\rho c_p(d\theta/\partial z) \qquad 12.17$$

According to several analyses on corn (*Zea mays*) the application of this formula to measured profiles of net radiation, temperature and humidity yields values of K which decrease exponentially with height as predicted by equation 12.15 and in some reports the exponential relation fits determinations of K almost perfectly from the top of the canopy down to the soil surface.[146] In a recent very careful analysis however,[64] the value of K in corn did not follow the simple exponential law but decreased faster than an exponential law would predict between h and $2h/3$, increased between $2h/3$ and $h/3$ and then decreased towards the soil surface (Fig. 12.11). This anomalous behaviour may be a consequence of the temperature profile in the crop: temperature decreased by 6°C between the soil surface and the top of the canopy and it is possible that turbulent transfer in this layer was enhanced by the effects of free convection.

Figure 12.11 shows that the error in K was about 10–20% in the top half of the canopy and larger lower down. If a similar error is involved in determining the gradient of temperature, humidity and CO_2, it would be difficult to measure the corresponding fluxes to better than ±25%. This figure implies that across any layer within the canopy where the flux is changing by 20%, say from 10 units to 8, the error in estimating the flux divergence of 2 units will be ±2·3 units, i.e. about 115%. Micrometeorologists have tended to minimize the size of this error which is an inevitable consequence of trying to estimate the small difference between two large quantities both of which have a substantial error. It is rare to find estimates of flux published with any indication of confidence limits.

Figure 12.12 shows an example of the analysis of vertical profiles of carbon dioxide flux in a corn canopy throughout a summer day.[55] The hatched areas represent the section of the canopy where the flux is upwards, i.e. below $h/2$ during most of the day but extending to the whole canopy after 1800. Corresponding profiles of flux divergence, expressed in units of g CO_2 per m^2 leaf per hour, show a rather irregular diurnal

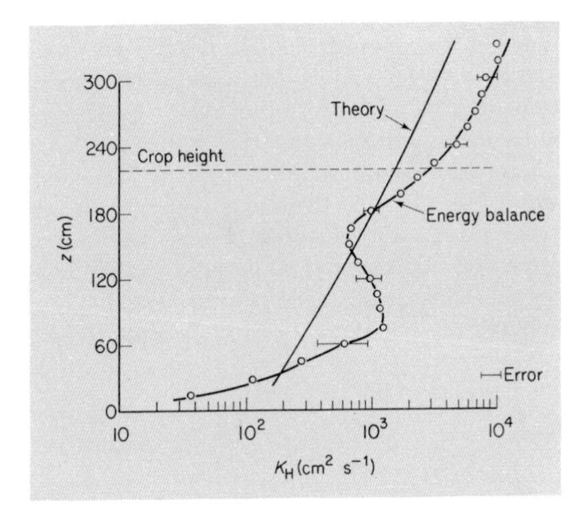

Fig. 12.11 Change of transfer coefficient for heat K_H in a field of corn (*Zea mays*) from Bowen ratio measurements at Ellis Hollow, Ithaca on 15 August 1968 (from Lemon[64]).

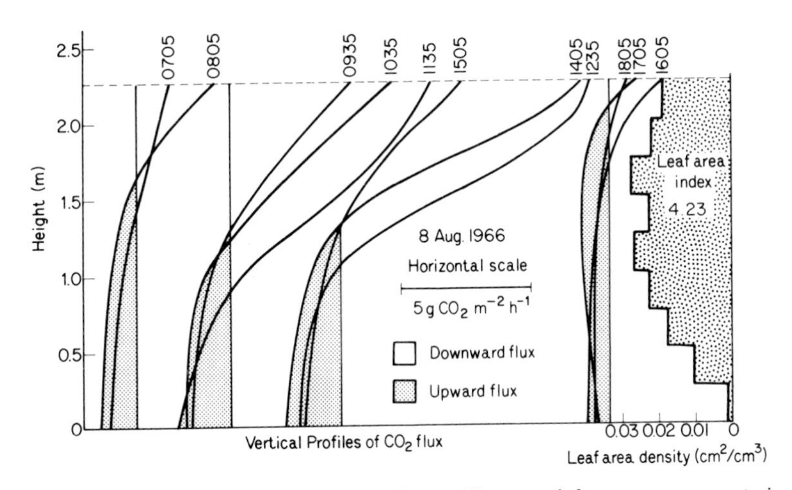

Fig. 12.12 Fluxes of CO_2 in a stand of corn (*Zea mays*) from measurements in Tokyo, Japan. The hatched areas represent heights where there is a net output of CO_2 by respiration, and above these areas the canopy acts as a sink of CO_2 (from Inoue *et al.*[55]).

change in the distribution of CO_2 sinks which extend down to the soil surface until 1700. If an error of $\pm 100\%$ is assigned to each point, the lack of consistency between the profiles can be ascribed to experimental error and may conceal more uniform behaviour in the real canopy.

The energy balance method of determining K yields as a by-product the fluxes of water vapour and sensible heat in the canopy, and an example is shown in Fig. 12.13. If the rate at which water is transpired in a layer and the relevant features of microclimate are known, it is possible to calculate the stomatal resistance r_s. When the mean temperature difference between leaves and air has been measured at different heights, it is possible to estimate the mean aerodynamic resistance r_H and the mean

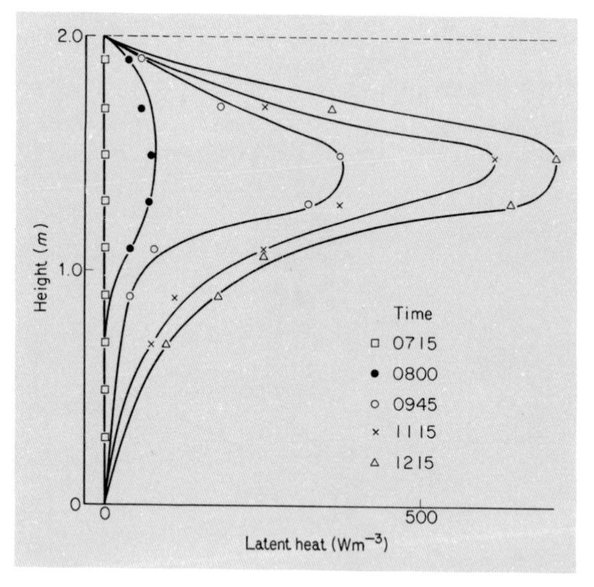

Fig. 12.13 Fluxes of water vapour in a stand of bulrush millet (*Pennisetum typhoides*) expressed in terms of a latent height flux; from measurements at Katherine, Australia, on 29 March 1963 (from Begg *et al.*[7]).

stomatal resistance r_s as a function of height. Japanese workers prefer to use the 'foliage exchange velocity' $D_f = 1/r_H$ and the 'stomatal exchange velocity' $D_s = 1/r_s$. They have shown[146] that in a corn crop r_H increases from about 0·3 s cm^{-1} near the top of the canopy to more than 1·0 s cm^{-1} at $z = h/2$ whereas r_s increases from a minimum of about 2 s cm^{-1} in the highest leaves to about 10 s cm^{-1} in the lowest leaves. These values are consistent with estimates of r_H from Nusselt numbers and with laboratory determinations of stomatal resistance using porometers.

The synthesis of profiles

The shape of profiles in plant stands can be simulated by reversing the analysis of the last section. Several stand models described in the literature follow the same procedure:

(i) the distribution of radiant energy in the canopy is expressed as a function of cumulative leaf area index;
(ii) the net radiation absorbed by each leaf is partitioned between sensible and latent heat;
(iii) the resistance between each layer of leaves is estimated as $\int (dz/K)$ where K is the turbulent diffusion coefficient;
(iv) the increase of the relevant potentials from the lowest layer to the second lowest layer is calculated from the product of flux and resistance. This procedure is repeated layer by layer till the top of the canopy is reached.

In the models described by Philip[105] and Cowan,[27] the individual leaf layers were made infinitely thin so that differential equations could be applied to the system. Philip assumed intuitively that the resistance r_H or r_V of individual leaves was inversely proportional to K at each level in the canopy and he neglected resistance r_s. The profiles he predicted were therefore valid only for foliage wetted by rain or dew when the stomatal resistance was negligible in comparison with the aerodynamic resistance, an unusual condition. Cowan used a stomatal resistance with the same value throughout the canopy, but Waggoner and Reifsnyder[153], using a finite difference method, allowed r_s to increase with decreasing irradiance.

The process of simulating microclimates from calculated fluxes is a challenging exercise but it is usually more straightforward to measure microclimates than to derive them from models. Micrometeorologists are more interested in estimating fluxes from profile shapes than *vice versa*. The main limitation of microclimatic models is ignorance about spatial changes of r_s in a canopy which have seldom been measured with precision. The importance of these changes is illustrated by Fig. 12.4 for which the evaporation from a stand of barley was calculated on the basis of three assumptions: (a) microclimate and r_s invariant with height, (b) r_s invariant; temperature, vapour pressure and wind varying as measured, (*c*) r_s estimated as a function of the irradiance; T, e and u as measured. The measured transpiration rate of 0·37 mm per hour was close to estimate (c) but was much smaller than (a) or (b). As a generaliza-

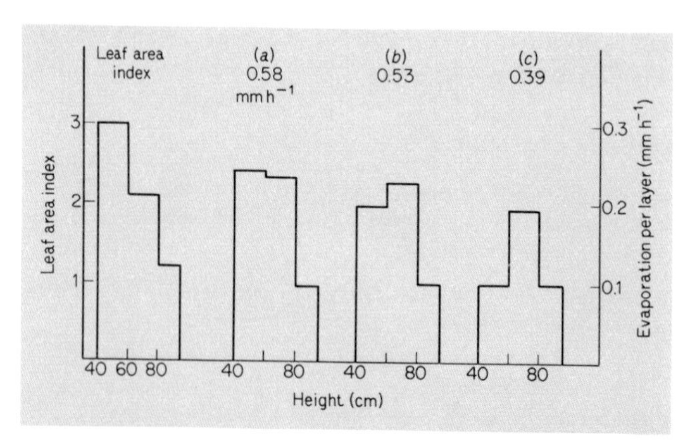

Fig. 12.14 Calculated transpiration from a stand of barley in three layers: 40 to 60, 60 to 80 and 80 to 95 cm, 23 July 1963.

(**a**) with T, e, u, r_s constant;
(**b**) with T, e, u varying with height as measured and r_s constant;
(**c**) with T, e, u and r_s all varying with height.

Figures above each section are total transpiration rates.

tion from this result, the exchange of heat and water vapour between a canopy and the air above it depends much more critically on the behaviour of the stomata than on the structure of the microclimate.

APPLICATIONS

This final chapter has reviewed progress in measuring and analysing the physical environment of crops to determine how transpiration and photosynthesis rates are governed by physical and physiological circumstances. As a guide to the extensive literature on the subject, Table 12.2 was compiled to summarize some of the more recent work on nine agricultural crops commonly grown in temperate and tropical climates.

The application of environmental physics to improve the yield of crops can take several forms: creating a better environment for growth; changing the response of a species to its environment; or selecting species which give a better yield in a given environment. Among many attempts to improve the environment of crops by better husbandry, the development of the Penman formula and its use to calculate irrigation needs has been outstandingly successful. The formula provides a convenient method of calculating the water needs of a crop from climatological measurements so that, when equipment and manpower are available, irrigation can be applied in the right amount at the right time.

Models for describing the much more complex process of crop growth have so far contributed relatively little to agricultural practice. Their main function has been to synthesize information about growth in the laboratory and in the field. To build more useful models, information is needed about the effects of temperature, water status and other physical factors on the decline in maximum photosynthesis rates as leaves age; on the distribution of assimilates; and on the relation between respiration and growth rates. Progress will depend on the extent to which biochemical principles can be incorporated in models of crop growth along with physical and physiological concepts which have been integrated already.

Many methods of modifying the physical environment have been practised for centuries: irrigation; planting tree shelter belts to reduce windspeed or to trap snow; increasing soil temperatures in spring by dusting with soot or by planting on a slope to intercept more radiation. Apart from the application of water, there seems little prospect of increasing the productivity of field crops substantially by modifying their physical environment. Uneconomic amounts of energy are needed to increase soil or air temperatures significantly or to supplement natural sources of radiation. Greater success may be achieved by changing the response of crop plants to their environment, for example, by increasing rates of photosynthesis in bright light or rates of leaf expansion at low temperatures.

For many years the study of crop–weather relationships was a sterile statistical exercise involving the application of multiple regression analysis to yields and climatological measurements. The subject is now developing rapidly and has reached a much more exciting phase in which the environmental physicist can play an important part. His expertise is needed for the design of instruments, the analysis and interpretation of measurements, the construction of models that simulate the behaviour of real crops, and the manipulation of the environment to improve both quality and yield.

Tables

Table A.1 Système International (SI) units with c.g.s. and British equivalents

Quantity	Dimensions	SI	c.g.s.	British
Length	L	1 m	= 10^2 cm	= 3·281 ft
Area	L^2	1 m^2	= 10^4 cm^2	= 10·72 ft^2
Volume	L^3	1 m^3	= 10^6 cm^3	= 35·31 ft^3
Mass	M	1 kg	= 10^3 g	= 2·205 lb
Density	$M L^{-3}$	1 kg m^{-3}	= 10^{-3} g cm^{-3}	= $6{\cdot}24 \times 10^{-2}$ lb ft^{-3}
Time	T	1 s (or min, h, etc)	= 1 s	= $2{\cdot}778 \times 10^{-4}$ h
Velocity	$L T^{-1}$	1 m s^{-1}	= 10^2 cm s^{-1}	= 3·281 ft s^{-1}
Acceleration	$L T^{-2}$	1 m s^{-2}	= 10^2 cm s^{-2}	= 3·281 ft s^{-2}
Force	$M L T^{-2}$	1 kg m s^{-2} = 1 N (Newton)	= 10^5 g cm^{-1} s^{-2} = 10^5 dynes	= 0·224 lb f (lb force)
Pressure	$M L^{-1} T^{-2}$	1 kg m^{-1} s^{-2} = 1 N m^{-2} (Pascal)	= 10 g cm^{-1} s^{-2} = 10^{-2} mbar	= 0·021 lb f ft^{-2}
Work, energy	$M L^2 T^{-2}$	1 kg m^2 s^{-2} = 1 J (Joule)	= 10^7 g cm^2 s^{-2} = 10^7 ergs	= 0·738 ft lb f
Power	$M L^2 T^{-3}$	1 kg m^2 s^{-3} = 1 W (Watt)	= 10^7 g cm^2 s^{-3} = 10^7 ergs s^{-1}	= 0·738 ft lb f s^{-1}
				= $1{\cdot}34 \times 10^{-3}$ hp
Dynamic viscosity	$M L^{-1} T^{-1}$	1 N s m^{-2}	= 10 Poise	= 0·672 lb ft^{-1} s^{-1}
Kinematic viscosity	$L^2 T^{-1}$	1 m^2 s^{-1}	= 10^4 Stokes	= 10·7 ft^2 s^{-1}
Temperature	θ	1 °C (or 1 K)	= 1 °C (or 1 K)	= 1·8 °F
Heat energy	H (or $M L^2 T^{-2}$)	1 J	= 0·2388 cal	= $9{\cdot}47 \times 10^{-4}$ BTU
Heat or radiation flux	$H T^{-1}$	1 W	= 0·2388 cal s^{-1}	= 3·412 BTU h^{-1}
Heat flux density	$H L^{-2} T^{-1}$	1 W m^{-2}	= $2{\cdot}388 \times 10^{-5}$ cal cm^{-2} s^{-1}	= 1·02 BTU ft^{-2} h^{-1}
Latent heat	$H M^{-1}$	1 J kg^{-1}	= $2{\cdot}388 \times 10^{-4}$ cal g^{-1}	= $4{\cdot}20 \times 10^{-4}$ BTU lb^{-1}
Specific heat	$H M^{-1} \theta^{-1}$	1 J kg^{-1} $°C^{-1}$	= $2{\cdot}388 \times 10^{-4}$ cal g^{-1} $°C^{-1}$	= $2{\cdot}388 \times 10^{-4}$ BTU lb^{-1} $°F^{-1}$
Thermal conductivity	$H L^{-1} \theta^{-1} T^{-1}$	1 W m^{-1} $°C^{-1}$	= $2{\cdot}388 \times 10^{-3}$ cal cm^{-2} s^{-1}	= 6·93 BTU ft^{-2} h^{-1} $°F^{-1}$ in.
Thermal diffusivity (and other diffusion coefficients)	$L^2 T^{-1}$	1 m^2 s^{-1}	= 10^4 cm^2 s^{-1}	= 10·8 ft^2 s^{-1}

Table A.2 Properties of air, water vapour and CO_2 (treated as constant between −5 and 45°C)

		Air	Water vapour	Carbon dioxide
Specific heat	$(J\ g^{-1}\ °C^{-1})$	1·01	1·88	0·85
Prandtl number	$Pr = \frac{\nu}{\kappa}$	$0{\cdot}70_5$	—	—
	$Pr^{0.67}$	0·79	—	—
	$Pr^{0.33}$	0·89	—	—
	$Pr^{0.25}$	0·92	—	—
Schmidt number	$Sc = \frac{\nu}{D}$	—	0·63	1·04
	$Sc^{0.67}$	—	0·74	1·02
	$Sc^{0.33}$	—	0·86	1·01
	$Sc^{0.25}$	—	0·89	1·01
Lewis number	$Le = \frac{\kappa}{D}$	—	0·89	1·48
	$Le^{0.67}$	—	0·93	1·32
	$Le^{0.33}$	—	0·96	1·14
	$Le^{0.25}$	—	0·97	1·11

Table A.3 Properties of air, water vapour and CO_2 (changing by less than 1% per °C)

	Temperature	Densities of air		Virtual temperature of air	Latent heat of vaporization of water		Thermal conductivity of air	Molecular diffusion coefficients of air			
Symbol	T	ρ_a	$\rho_{as}(T)$	T_v	λ	γ	k	κ	ν	D_V	D_C
Unit °C	K	kg m^{-3}		°C	J g^{-1}	mbar °C^{-1}	mW $m^{-1}K^{-1}$	cm^2 s^{-1}			
−5	268·2	1·316	1·314	−4·57	2513	0·643	24·0	0·183	0·129	0·205	0·124
0	273·2	1·292	1·289	0·64	2501	0·646	24·3	0·189	0·133	0·212	0·129
5	278·2	1·269	1·265	5·92	2489	0·649	24·6	0·195	0·137	0·220	0·133
10	283·2	1·246	1·240	11·32	2477	0·652	25·0	0·202	0·142	0·227	0·138
15	288·2	1·225	1·217	16·87	2465	0·655	25·3	0·208	0·146	0·234	0·142
20	293·2	1·204	1·194	22·62	2454	0·658	25·7	0·215	0·151	0·242	0·147
25	298·2	1·183	1·169	28·62	2442	0·662	26·0	0·222	0·155	0·249	0·151
30	303·2	1·164	1·145	34·97	2430	0·665	26·4	0·228	0·160	0·257	0·156
35	308·2	1·146	1·121	41·73	2418	0·668	26·7	0·235	0·164	0·264	0·160
40	313·2	1·128	1·096	49·03	2406	0·671	27·0	0·242	0·169	0·272	0·165
45	318·2	1·110	1·068	57·02	2394	0·675	27·4	0·249	0·174	0·280	0·170

ρ_a density of dry air
$\rho_{as}(T)$ density of air saturated with water vapour at temperature T
T_v virtual temperature of saturated air
λ latent heat of vaporization of water
γ $c_p/\lambda\varepsilon$—the 'psychrometer constant'

k thermal conductivity of dry air
κ thermal diffusivity of dry air
ν kinematic viscosity of dry air
D_V diffusion coefficient of water vapour in air
D_C diffusion coefficient of CO_2 in air

Table A.4 Quantities changing by more than 1% per °C

$e_s(T)$ saturation vapour pressure at temperature T (°C)
Δ change of saturation vapour pressure per °C, i.e. $\partial e_s/\partial T$
σT^4 full radiation at temperature T (K)
$4\sigma T^3$ change of full radiation per K

Note that the quantities Δ and $4\sigma T^3$ can be used as mean differences to interpolate between the tabulated values of e and σT^4 respectively.

T (°C)	T (K)	$e_s(T)$ (mbar)	$\Delta(T)$ mbar °C^{-1}	σT^4 W m^{-2}	$4\sigma T^3$ W m^{-2} K^{-1}
−5	268·2	4·21	0·32	293·4	4·4
−4	269·2	4·55	0·34	297·8	4·4
−3	270·2	4·90	0·37	302·2	4·5
−2	271·2	5·28	0·39	306·7	4·5
−1	272·2	5·68	0·42	311·3	4·6
0	273·2	6·11	0·45	315·9	4·6
1	274·2	6·57	0·48	320·5	4·7
2	275·2	7·05	0·51	325·2	4·7
3	276·2	7·58	0·54	330·0	4·8
4	277·2	8·13	0·57	334·8	4·8
5	278·2	8·72	0·61	339·6	4·9
6	279·2	9·35	0·65	344·5	5·0
7	280·2	10·01	0·69	349·5	5·0
8	281·2	10·72	0·73	354·5	5·1
9	282·2	11·47	0·78	359·6	5·1
10	283·2	12·27	0·83	364·7	5·2
11	284·2	13·12	0·88	369·9	5·2
12	285·2	14·02	0·93	375·1	5·3
13	286·2	14·97	0·98	380·4	5·3
14	287·2	15·98	1·04	385·8	5·4
15	288·2	17·04	1·10	391·2	5·4
16	289·2	18·17	1·17	396·6	5·5
17	290·2	19·37	1·23	402·1	5·6
18	291·2	20·63	1·30	407·7	5·6
19	292·2	21·96	1·37	413·3	5·7
20	293·2	23·37	1·45	419·0	5·7
21	294·2	24·86	1·53	424·8	5·8
22	295·2	26·43	1·62	430·6	5·8
23	296·2	28·09	1·70	436·4	5·9
24	297·2	29·83	1·79	442·4	6·0
25	298·2	31·67	1·89	448·3	6·0

Table A.4—*continued*

T (°C)	(K)	$e_s(T)$ (mbar)	$\Delta(T)$ mbar°C^{-1}	σT^4 W m^{-2}	$4\sigma T^3$ W m^{-2} K^{-1}
26	299·2	33·61	1·99	454·4	6·1
27	300·2	35·65	2·10	460·5	6·2
28	301·2	37·80	2·21	466·7	6·2
29	302·2	40·06	2·32	472·9	6·3
30	303·2	42·43	2·44	479·2	6·3
31	304·2	44·93	2·57	485·5	6·4
32	305·2	47·55	2·69	492·0	6·5
33	306·2	50·31	2·83	498·4	6·6
34	307·2	53·20	2·97	505·0	6·6
35	308·2	56·24	3·12	511·6	6·7
36	309·2	59·42	3·27	518·3	6·7
37	310·2	62·76	3·43	525·0	6·8
38	311·2	66·26	3·57	531·8	6·9
39	312·2	69·93	3·76	538·7	6·9
40	313·2	73·78	3·94	545·6	7·0
41	314·2	77·80	4·13	552·6	7·1
42	315·2	82·02	4·32	559·7	7·1
43	316·2	86·42	4·52	566·8	7·2
44	317·2	91·03	4·73	574·0	7·3
45	318·2	95·86	4·94	581·3	7·3

Table A.5 Nusselt numbers for air
(a) Forced convection

Shape	Case	Range of Re	Nu
(1) Flat plates			
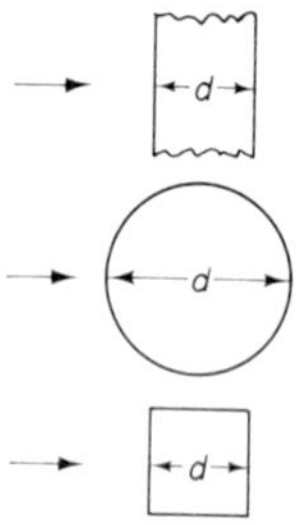	Streamline flow	$<2 \times 10^4$	$0{\cdot}60\ Re^{0{\cdot}5}$
	Turbulent flow	$>2 \times 10^4$	$0{\cdot}032\ Re^{0{\cdot}8}$
(2) Cylinders			
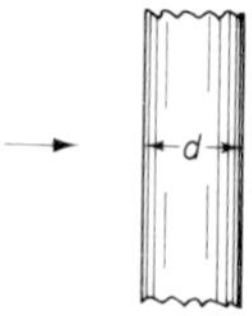	Narrow range of Reynolds numbers	1–4	$0{\cdot}89\ Re^{0.33}$
		4–40	$0{\cdot}82\ Re^{0.39}$
		$40 - 4 \times 10^3$	$0{\cdot}62\ Re^{0.47}$
		$4 \times 10^3 - 4 \times 10^4$	$0{\cdot}17\ Re^{0.62}$
		$4 \times 10^4 - 4 \times 10^5$	$0{\cdot}024\ Re^{0.81}$
		or	
	Wide range of Reynolds numbers	$10^{-1} - 10^3$	$0{\cdot}32 + 0{\cdot}51\ Re^{0.52}$
		$10^3 - 5 \times 10^4$	$0{\cdot}24\ Re^{0{\cdot}60}$
(3) Spheres			
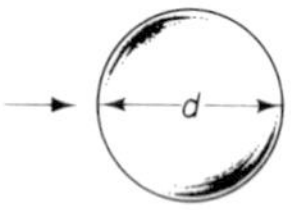		0–300	$2 + 0{\cdot}54\ Re^{0{\cdot}5}$
		$50 - 1{\cdot}5 \times 10^5$	$0{\cdot}34\ Re^{0{\cdot}6}$

Notes
(i) Arrows show direction of airflow
(ii) d is characteristic dimension; take width of a long crosswind strut as shown or mean side for a rectangle whose width and length are comparable
(iii) To find corresponding Sherwood numbers multiply Nu by $Le^{0{\cdot}33}$ (see values in Table A.1)
(iv) Sources—Ede,[38] Fishenden and Saunders,[40] Bird, Stewart and Lightfoot.[8]

Table A.5—(*continued*)
(b) Free convection

Shape and relative temperature	Range: Laminar flow	Range: Turbulent flow	Nu
(1) Horizontal flat plates or cylinders			
(i) 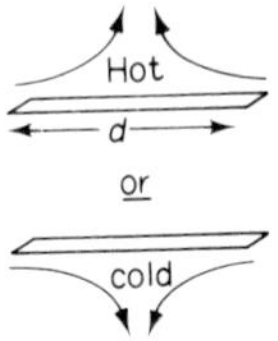 Hot (d) or cold	Gr < 10^5		$0{\cdot}50\ Gr^{0{\cdot}25}$
		Gr > 10^5	$0{\cdot}13\ Gr^{0{\cdot}33}$
(ii) 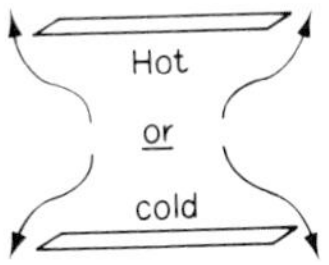 Hot or cold		Arrangement not conducive to turbulence	$0{\cdot}23\ Gr^{0{\cdot}25}$
(iii)  Hot or cold (d)	10^4 < Gr < 10^9		$0{\cdot}48\ Gr^{0{\cdot}25}$
		Gr > 10^9	$0{\cdot}09\ Gr^{0{\cdot}33}$
(2) Vertical flat plates or cylinders: 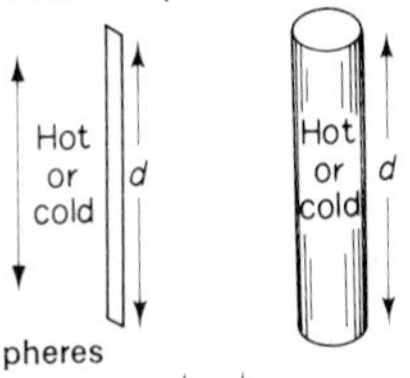 Hot or cold (d)	10^4 < Gr < 10^9		$0{\cdot}58\ Gr^{0{\cdot}25}$
		10^9 < Gr < 10^{12}	$0{\cdot}11\ Gr^{0{\cdot}33}$
(3) Spheres: 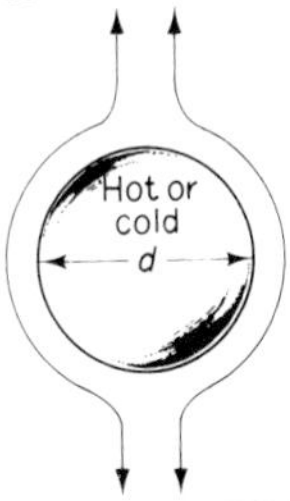 Hot or cold (d)	$Gr^{0{\cdot}25}$ < 220		$2+0{\cdot}54\ Gr^{0{\cdot}25}$

Notes
(i) Arrows indicate direction of air circulation
(ii) d is characteristic dimensions for calculation of Gr: take height for vertical plate and average chord for horizontal plate
(iii) To find corresponding Sherwood numbers, multiply Nu by $Le^{0{\cdot}25}$ for laminar flow or turbulent flow (see values in Table A.1)
(iv) Sources—Ede,[38] Fishenden and Saunders,[40] Bird, Stewart and Lightfoot.[8]

Bibliography

GENERAL INTRODUCTORY TEXT BOOKS

Energy Exchange in the Biosphere, D. M. GATES. Harper and Row, New York, 1962.
Introduction to radiation and heat exchange.

Physical Climatology, W. D. SELLERS. University of Chicago Press, 1965.
Discussion of radiation, heat and water balances.

The Climate near the Ground, R. GEIGER. Harvard University Press, Cambridge, Mass., 1965.
A review and discussion of more than 1200 papers in the literature of microclimatology.

Agricultural Physics, C. W. ROSE. Pergamon Press, Oxford, 1966.
Special emphasis on the physics of soil water.

Applied Climatology, J. F. GRIFFITHS. Oxford University Press, 1966.
The application of climatological measurements to human activities.

Weather and Life, W. P. LOWRY. Academic Press, New York, 1969.
The biometeorology of plants, animals and human beings. Numerical problems.

MORE SPECIALIZED TEXT BOOKS

Radiation

Solar Radiation, ed. N. ROBINSON. Elsevier, Amsterdam, 1966.
Detailed treatment of the transmission of solar radiation in the atmosphere.

Radiation in the Atmosphere, K. YA. KONDRATYEV. Academic Press, New York, 1969.
Very comprehensive treatment of short and long wave radiation.

Heat Transfer

An Introduction to Heat Transfer, A. J. EDE. Pergamon Press, Oxford, 1967.
Clear presentation using SI units; useful tables.

Principles of Heat Transfer, F. KREITH. International Textbook Co., Scranton, Pa., U.S.A., 1958.
Emphasizes engineering applications.

Physics of Plant Environment, ed. W. R. VAN WIJK. North-Holland Publishing Company, Amsterdam, 1963.
Mainly concerned with heat transfer in the soil, free atmosphere, and glasshouses.

Micrometeorology

Micrometeorology, O. G. SUTTON. McGraw Hill, New York, 1953.
Predominantly mathematical treatment.
Turbulent Transfer in the Lower Atmosphere, C. H. B. PRIESTLEY. University of Chicago Press, 1959.
Short but comprehensive survey of principles and of measurements before 1959.
The Structure of Atmospheric Turbulence, J. L. LUMLEY and H. A. PANOFSKY. Wiley, New York, 1964.
Agricultural Meteorology, ed. P. E. WAGGONER. American Meteorological Society, Boston, 1965.
Contains a number of useful review articles, e.g. on radiation (Gates), atmospheric turbulence (Webb), etc.
Descriptive Micrometeorology, R. E. MUNN. Academic Press, New York, 1966.
Good reference source.

Animal Heat Balance

Man in a Cold Environment, R. C. BURTON and O. G. EDHOLM. Edward Arnold, London, 1955.
Analysis of human heat balance in relation to physiology.
The Fire of Life, M. KLEIBER. J. Wiley, New York, 1961.
The history and methodology of energy metabolism.
Thermobiology, ed. A. H. ROSE. Academic Press, London, 1967.
Chapters on thermal relations of plants, insects, animals and man.
The Energy Metabolism of Ruminants, K. L. BLAXTER. Hutchinson, London, 1967.
Links biochemistry of energy production to the physics of dissipation.
The Climatic Physiology of the Pig, L. E. MOUNT. Edward Arnold, London, 1968.
Analysis of the heat balance of pigs in relation to their physiology.

Crop Micrometeorology

Relevant chapters will be found in the proceedings of conferences published in the following volumes:

Environmental Control of Plant Growth, ed. L. T. EVANS. Academic Press, New York, 1963.
Functioning of Terrestrial Ecosystems, ed. F. E. ECKARDT. UNESCO, Paris, 1968.
Physiological Aspects of Crop Yield, ed. J. D. EASTIN. American Society of Agronomy, Madison, Wis., U.S.A., 1969.

Plant and Soil Water Relations

Plant-Water Relationships, R. O. SLATYER. Academic Press, New York, 1967.
Comprehensive review of physical and physiological aspects.

Soil and Water, D. HILLEL. Academic Press, New York, 1971.
The physics of water movement in soils and its biological implications.

Particulate Diffusion

Atmospheric Diffusion, F. PASQUILL. Van Nostrand, New York, 1962.

Airborne Microbes, ed. P. H. GREGORY and J. L. MONTEITH. Seventeenth Symp. Soc. Gen. Microbiol., Cambridge University Press, 1967.
Several chapters are concerned with the dispersal and deposition of microbes in relation to the behaviour of the atmosphere.

Instrumentation

Instruments for Micrometeorology (IBP Handbook No. 22), ed. J. L. MONTEITH. Blackwells Scientific Publications, Oxford, 1972.
Contains specifications of instruments, references to the literature (1960–1970) and a comprehensive bibliography.

References

1. ACOCK, B., THORNLEY, J. H. M. and WARREN WILSON, J. (1970). Spatial variation of light in the canopy. In *Prediction and Measurement of Photosynthetic Productivity*, SETLIK, I. IBP/PP Technical Meeting, Trebon, Pudoc, Wageningen.
2. ANDERSON, M. C. (1966). Stand structure and light penetration. II. A theoretical analysis. *J. appl. Ecol.*, **3**, 41.
3. ANDERSON, M. C. and DENMEAD, O. T. (1969). Short wave radiation on inclined surfaces in model plant communities. *Agron. J.*, **61**, 867.
4. BARRY, R. G. and CHAMBERS, R. E. (1966). A preliminary map of summer albedo over England and Wales. *Q. Jl R. met. Soc.*, **92**, 543.
5. BARTLETT, P. N. and GATES, D. M. (1967). The energy budget of a lizard on a tree trunk. *Ecology*, **48**, 315.
6. BAUMGARTNER, A. (1953). Das Eindringen des Lichtes in den Boden. *Forstwiss. ZentBl.*, **72**, 172.
7. BEGG, J. E., BIERHUIZEN, J. F., LEMON, E. R., MISRA, D., SLATYER, R. O. and STERN, W. R. (1964). Diurnal energy and water exchanges in bulrush millet. *Agric. Meteorol.*, **1**, 294.
8. BIRD, R. B., STEWART, W. E. and LIGHTFOOT, E. N. (1960). *Transport Phenomena*. John Wiley, New York.
9. BIRKEBAK, R. C. (1967). Heat transfer in biological systems. *Int. Rev. Gen. & Exp. Zool.*, **2**, 269.
10. BISCOE, P. V. (1969). *Stomata and the plant environment*. Ph.D. thesis, University of Nottingham.
11. BLACK, T. A., TANNER, C. B. and GARDNER, W. R. (1970). Evapotranspiration from a snap bean crop. *Agron. J.*, **62**, 66.
12. BLAXTER, K. L. (1967). *The Energy Metabolism of Ruminants*. 2nd edition. Hutchinson, London.
13. BOLIN, B. and BISCHOF, W. (1970). Variation of the carbon dioxide content of the atmosphere in the northern hemisphere. *Tellus*, **21**, 431.
14. BOWERS, S. A. and HANKS, R. D. (1965). Reflection of radiant energy from soils. *Soil Sci.*, **100**, 130.
15. BRADLEY, E. F. (1968). A micrometeorological study of velocity profiles and surface drag in the region modified by a change in surface roughness. *Q. Jl R. met. Soc.*, **94**, 361.
16. BRENCHLEY, G. H. (1968). Aerial photography for the study of plant disease. *A. Rev. Phytopathol.*, **6**, 1.

17. BROWN, K. W. and COVEY, W. (1966). The energy budget evaluation of the micrometeorological transfer processes within a corn field. *Agric. Meteorol.*, **3**, 71.
18. BRUNT, D. (1939). *Physical and Dynamical Meteorology*. Cambridge University Press.
19. BUETTNER, K. J. K. and KERN, C. D. (1965). The determination of infra-red emissivities of terrestrial surfaces. *J. geophys. Res.*, **70**, 1329.
20. BURTON, A. C. and EDHOLM, O. G. (1955). *Man in a Cold Environment*. Edward Arnold, London.
21. CHAMBERLAIN, A. C. (1953). Experiments on the deposition of Iodine 131 vapour onto surfaces from an airstream. *Phil. Mag.*, **44**, 1145.
22. CHAMBERLAIN, A. C. (1966). Transport of gases to and from grass and grass-like surfaces. *Proc. R. Soc. A*, **290**, 236.
23. CHRENKO, F. A. and PUGH, L. G. Ç. E. (1961). The contribution of solar radiation to the thermal environment of man in Antarctica. *Proc. R. Soc. B*, **155**, 243.
24. CHURCH, N. S. (1960). Heat loss and the body temperatures of flying insects. *J. exp. Biol.*, **37**, 171.
25. CLAPPERTON, J. L., JOYCE, J. P. and BLAXTER, K. L. (1965). Estimates of the contribution of solar radiation to thermal exchanges of sheep at a latitude of 55°N. *J. agric. Sci., Camb.*, **64**, 37.
26. COLWELL, R. N. (1968). Remote sensing of natural resources. *Scient. Am.*, **218**, 54.
27. COWAN, I. R. (1968). Mass, heat and momentum exchange between stands of plants and their atmospheric environment. *Q. Jl R. met. Soc.*, **94**, 523.
28. DAVIES, J. A. and BUTTIMOR, P. H. (1969). Reflection coefficients, heating coefficients and net radiation at Simcoe, S. Ontario. *Agric. Meteorol.*, **6**, 373.
29. DEACON, E. L. (1957). Wind profiles and the shearing stress—an anomaly resolved. *Q. Jl R. met. Soc.*, **83**, 537.
30. DEACON, E. L. (1969). Physical processes near the surface of the earth. In *World Survey of Climatology*, Vol. 2, *General Climatology*, LANDSBERG, H. E. Elsevier, Amsterdam.
31. DENMEAD, O. T. (1969). Comparative micrometeorology of a wheat field and a forest of *Pinus radiata*. *Agric. Meteorol.*, **6**, 357.
32. DIGBY, P. S. B. (1955). Factors affecting the temperature excess of insects in sunshine. *J. exp. Biol.*, **32**, 279.
33. DINES, W. H. and DINES, L. H. G. (1927). Monthly mean value of radiation from various parts of the sky at Benson, Oxfordshire. *Mem R. met. Soc.*, **2**, No. 11.
34. DOGNIAUX, R. (1954). Étude du climat de la radiation en Belgique. *Inst. Roy. Met. Belgique*, Contrib. No. **18**.
35. DONEY, J. M. (1963). The effects of exposure in Blackface sheep. *J. agric. Sci., Camb.*, **60**, 267.
36. DYER, A. J. and HICKS, B. B. (1970). Flux-gradient relationships in the constant flux layer. *Q. Jl R. met. Soc.*, **96**, 715.
37. ECKARDT, F. E., HEIM, G., METHY, M., SAUGIER, B. and SAUVEZON, R. (1971). Fonctionnement d'un ecosystème au nivau de la production primaire mesures effectuées dans une culture d'*Helianthus annuus*. *Oecol. Plant.*, **6**, 51.
38. EDE, A. J. (1967). *An Introduction to Heat Transfer Principles and Calculations*. Pergamon Press, Oxford.

39. ELSASSER, W. M. (1942). Heat transfer by infra-red radiation in the atmosphere. *Harv. met. Stud.*, **6**, 107.
40. FISHENDEN, M. and SAUNDERS, O. A. (1950). *An Introduction to Heat Transfer*. Clarendon Press, Oxford.
41. FLEISCHER, R. (1955). Der Jahresgang der Strahlungsbilanz solvie ihrer langund kurzwelligen Komponenten. *Ber. dt. Wetterd., Bad Kissingen*, No. **82**.
42. FRASER, A. I. (1962). *Wind tunnel studies of the forces acting on the crowns of small trees*. Rep. Forest Res., Lond.
43. GABRIELSEN, E. K. (1948). Effects of different chlorophyll concentrations on photosynthesis in foliage leaves. *Physiologia Pl.*, **1**, 5.
44. GARNIER, B. J. and OHMURA, A. (1968). A method of calculating the direct short wave radiation income of slopes. *J. appl. Meteorol.*, **7**, 706.
45. GATES, D. M. (1962). *Energy Exchange in the Biosphere*. Harper and Row, New York.
46. GATES, D. M., KEEGAN, H. N., SCHLETER, J. C. and WEIDNER, V. R. (1965). Spectral properties of plants. *Appl. Opt.*, **4**, 11.
47. HAMMEL, H. T. (1955). Thermal properties of fur. *Am. J. Physiol.*, **182**, 369.
48. HATCH, T. (1963). Assessment of heat stress. In *Temperature, its Measurement and Control in Science and Industry*, ed. HERZFELD, C. M. Reinhold Publishing Corporation, New York.
49. HEATH, O. V. S. (1969). *The Physiological Aspects of Photosynthesis*. Heinemann Educational Books, London.
50. HEMMINGSEN, A. M. (1960). Energy metabolism as related to body size and respiratory surfaces. *Rep. Steno meml Hosp., Copenhagen*, **9**, part 2, 7.
51. HUNT, L. A., IMPENS, I. I. and LEMON, E. R. (1968). Estimates of the diffusion resistance of some large sunflower plants in the field. *Pl. Physiol., Lancaster*, **43**, 522.
52. HUTCHINSON, J. C. D. and BROWN, G. D. (1969). Penetrance of cattle coats by radiation. *J. appl. Physiol.*, **26**, 454.
53. IDSO, S. B., JACKSON, R. D., EHRLER, W. L. and MITCHELL, S. T. (1969). A method for determination of infrared emittance of leaves. *Ecology*, **50**, 899.
54. IMPENS, I. (1965). *Experimentele Studie van de Thysische en Biologische Aspektera van de Transpiratie*. Rykslandbouwhogeschool, Gent.
55. INOUE, E., UCHIJIMA, Z., UDAGAWA, T., HORIE, T. and KOBYASHI, K. (1968). CO_2 flux within and above a corn plant canopy. *J. agric. Met., Tokyo*, **23**, 165.
56. JOYCE, J. L., BLAXTER, K. L. and PARK, C. (1966). The effect of natural outdoor environments on the energy requirements of sheep. *Res. vet. Sci.*, **7**, 342.
57. KALMA, J. D. (1970). *Some aspects of the water balance of an irrigated orange plantation*. Special Publication, Volcani Institute of Agricultural Research, Israel.
58. KLEIBER, M. (1965). Metabolic body size. In *Energy Metabolism*, BLAXTER, K. L. Academic Press, London.
59. KONDRATYEV, K. J. and MANOLOVA, M. P. (1960). The radiation balance of slopes. *Sol. Energy*, **4**, 14.
60. KONDRATYEV, K. J. and NIKOLSKY, G. A. (1970). Solar radiation and solar activity. *Q. Jl R. met. Soc.*, **96**, 509.

61. KREITH, F. (1958). *Principles of Heat Transfer*. International Text Book Co., Scranton, Pa., U.S.A.
62. KUROIWA, S. (1968). A new calculation method for total photosynthesis of a plant community. In *Functioning of Terrestrial Ecosystems*, ECKARDT, F. E. UNESCO, Paris.
63. LEMON, E. R. (1960). Photosynthesis under field conditions. II. An aerodynamic method for determining the turbulent carbon dioxide exchange between the atmosphere and a corn field. *Agron. J.*, **52**, 697.
64. LEMON, E. R. (1970). Mass and energy exchange between plant stands and environment. In *Prediction and Measurement of Photosynthetic Productivity*, SETLIK, I. IBP/PP Technical Meeting, Trebon, Pudoc, Wageningen.
65. LEMON, E. R. and WRIGHT, J. L. (1969). Photosynthesis under field conditions. X.A. Assessing sources and sinks of carbon dioxide in a corn crop using a momentum balance approach. *Agron. J.*, **61**, 408.
66. LENTZ, C. P. and HART, J. S. (1960). The effect of wind and moisture on heat loss through the fur of newborn caribou. *Can. J. Zool.*, **38**, 679.
67. LETTAU, H. H. (1969). Note on aerodynamic roughness—parameter estimation on the basis of roughness element description. *J. appl. Meteorol.*, **8**, 828.
68. LEWIS, H. E., FORSTER, A. R., MULLAN, B. J., COX, R. N. and CLARK, R. P. (1969). Aerodynamics of the human microenvironment. *Lancet*, 1273.
69. LONG, I. F., MONTEITH, J. L., PENMAN, H. L. and SZEICZ, G. (1964). The plant and its environment. *Met. Rdsch.*, **17**, 97.
70. LONG, I. F. and PENMAN, H. L. (1964). The micrometeorology of the potato crop. In *The Growth of the Potato*, IVINS, J. D. and MILTHORPE, F. L. Butterworths, London.
71. LUMB, F. E. (1964). The influence of cloud on hourly amounts of total solar radiation at the sea surface. *Q. Jl R. met. Soc.*, **90**, 43.
72. MCCULLOCH, J. S. G. and PENMAN, H. L. (1956). Heat flow in the soil. *Proc. 6th International Soil Science Congr.*, **1**, 275.
73. MACFARLANE, W. V., MORRIS, R. J. and HOWARD, B. (1956). Water economy of tropical merino sheep. *Nature, Lond.*, **178**, 304.
74. MAKI, T. (1969). On zero plane displacement and roughness length in the wind velocity profile over a corn canopy. *J. agric. Met., Tokyo*, **25**, 13.
75. MATTSSON, J. D. (1961). Microclimate observations in and above cultivated crops with special regard to temperature and relative humidity. *Lund Stud. Geogr., Ser. A*, No. **16**.
76. MATTSSON, J. O. (1966). The temperature climate of potato crops. *Lund Stud. Geogr., Ser. A*, No. **35**.
77. MEIDNER, H. and MANSFIELD, T. A. (1968). *Physiology of Stomata*. McGraw Hill, London.
78. MELLOR, R. S., SALISBURY, F. B. and RASCHKE, K. (1964). Leaf temperatures in controlled environments. *Planta*, **61**, 56.
79. MILTHORPE, F. L. and PENMAN, H. L. (1967). The diffusive conductivity of the stomata of wheat leaves. *J. exp. Bot.*, **18**, 422.
80. MITCHELL, D., WYNSHAM, C. H. and HODGSON, T. (1967). The selection of a biothermal radiometer. *J. scient. Instrum.*, **44**, 847.
81. MONTEITH, J. L. (1961). An empirical method for estimating long wave radiation exchanges in the British Isles. *Q. Jl R. met. Soc.*, **87**, 171.
82. MONTEITH, J. L. (1962). Measurement and interpretation of carbon dioxide fluxes in the field. *Neth. J. agric. Sci.*, **10**, 334.

83. MONTEITH, J. L. (1962). Attenuation of solar radiation: a climatological study. *Q. Jl R. met. Soc.*, **87**, 171.
84. MONTEITH, J. L. (1963). Dew: facts and fallacies. In *The Water Relations of Plants*, ed. RUTTER, A. J. and WHITEHEAD, F. H., Blackwell Scientific Publications, Oxford.
85. MONTEITH, J. L. (1963). Gas exchange in plant communities. In *Environmental Control of Plant Growth*, ed. EVANS, L. T. Academic Press, New York.
86. MONTEITH, J. L. (1964) Evaporation and environment. In *The State and Movement of Water in Living Organisms. 19th Symp. Soc. exp. Biol.*, 205.
87. MONTEITH, J. L. (1965). Radiation and crops. *Exp. Agric.*, **1**, 241.
88. MONTEITH, J. L. (1965). Light distribution and photosynthesis in field crops. *Ann. Bot.*, **29**, 17.
89. MONTEITH, J. L. (1969). Light interception and radiative exchange in crop stands. In *Physiological Aspects of Crop Yield*, ed. EASTIN, J. D. American Society of Agronomy, Madison, Wis.
90. MONTEITH, J. L. and SZEICZ, G. (1960). The carbon dioxide flux over a field of sugar beet. *Q. Jl R. met. Soc.*, **86**, 205.
91. MONTEITH, J. L. and SZEICZ, G. (1961). The radiation balance of bare soil and vegetation. *Q. Jl R. met. Soc.*, **87**, 159.
92. MONTEITH, J. L., SZEICZ, G. and YABUKI, K. (1964). Crop photosynthesis and the flux of carbon dioxide below the canopy. *J. appl. Ecol.*, **1**, 321.
93. MONTEITH, J. L., SZEICZ, G. and WAGGONER, P. E. (1965). The measurement and control of stomatal resistance in the field. *J. appl. Ecol.*, **2**, 345.
94. MOON, P. (1940). Proposed standard radiation curve. *J. Franklin Inst.*, **230**, 583.
95. MOUNT, L. E. (1968). *The Climatic Physiology of the Pig.* Edward Arnold, London.
96. NEWTON, J. E. and BLACKMAN, G. E. (1970). The penetration of solar radiation through leaf canopies of different structure. *Ann. Bot.*, **34**, 329.
97. NILSON, T. (1971). A theoretical analysis of the frequency of gaps in plant stands. *Agric. Meteorol.*, **8**, 25.
98. OGUNTOYINBO, J. S. (1970). Reflection coefficient of natural vegetation, crops and urban surfaces in Nigeria. *Q. Jl R. met. Soc.*, **96**, 430.
99. PARKHURST, D. F., DUNCAN, P. R., GATES, D. M. and KREITH, F. (1968). Wind-tunnel modelling of convection of heat between air and broad leaves of plants. *Agric. Meteorol.*, **5**, 33.
100. PARLANGE, J-Y., WAGGONER, P. E. and HEICHEL, G. H. (1971). Boundary layer resistance and temperature distribution on still and flapping leaves. *Pl. Physiol., Lancaster*, **48**, 437.
101. PAULSON, C. A. (1970). The mathematical representation of wind speed and temperature profiles in the unstable atmospheric surface layer. *J. appl. Meteorol.*, **9**, 857.
102. PENMAN, H. L. (1948). Natural evaporation from open water, bare soil and grass. *Proc. R. Soc. A*, **194**, 120.
103. PENMAN, H. L. and LONG, I. F. (1960). Weather in wheat: an essay in micrometeorology, *Q. Jl R. met. Soc.*, **86**, 16.
104. PENMAN, H. L. and SCHOFIELD, R. K. (1951). Some physical aspects of assimilation and transpiration. *Symp. Soc. exp. Biol.*, **5**, 115.

105. PHILIP, J. R. (1964). Sources and transfer processes in the air layer occupied by vegetation. *J. appl. Meteorol.*, **4**, 390.
106. PHILIP, J. R. (1966). Plant water relations: some physical aspects. *A. Rev. Pl. Physiol.*, **17**, 245.
107. PORTER, W. P. and GATES, D. M. (1969). Thermodynamic equilibria of animals with environment. *Ecol. Monogr.*, **39**, 245.
108. POWELL, R. W. (1940). Further experiments on the evaporation of water from saturated surfaces. *Trans. Instn chem. Engrs*, **18**, 36.
109. PRIESTLEY, C. H. B. (1959). *Atmospheric Turbulence*. University of Chicago Press.
110. PRIESTLEY, C. H. B. (1957). The heat balance of sheep standing in the sun. *Aust. J. agric. Res.*, **8**, 271.
111. RAINEY, R. C., WALOFF, Z. and BURNETT, G. F. (1957). *The behaviour of the Red Locust*. Anti-Locust Research Centre, London.
112. RAPP, G. M. (1970). Convective mass transfer and the coefficient of evaporative heat loss from the human skin. In *Physiological and Behavioral Temperature Regulation*, ed. HARDY, J. D., GAGGE, A. P. and STOLWIJK, J. A. J. C. C. Thomas, Illinois.
113. RASCHKE, K. (1956). Über die physikalischen Beziehungen zwischen Wärmeübergangszahl, Strahlungsaustausch, Temperatur und Transpiration eines Blattes. *Planta*, **48**, 200.
114. REEVE, J. E. (1960). Appendix to Inclined Point Quadrats by J. Warren Wilson. *New Phytol.*, **59**, 1.
115. RIDER, N. E. (1954). Evaporation from an oat field. *Q. Jl R. met. Soc.*, **80**, 198.
116. RIJKS, D. A. (1967). Water use by irrigated cotton in Sudan. I. Reflection of short wave radiation. *J. appl. Ecol.*, **4**, 561.
117. RIJKS, D. A. (1968). Water use by irrigated cotton in Sudan. II. Net radiation and soil heat flux. *J. appl. Ecol.*, **5**, 685.
118. ROSS, J. (1970). Mathematical models of photosynthesis in a plant stand. In *Prediction and Measurement of Photosynthetic Productivity*, SETLIK, I. IBP/PP Technical Meeting, Trebon, Pudoc, Wageningen.
119. SAUGIER, B. (1970). Transports turbulents de CO_2 et de vapeur d'eau au-dessus et à l'intérieur de la végétation. *Oecol. Plant.*, **5**, 179.
120. SCHLICHTING, H. (1960). *Boundary Layer Theory*. 4th edition. McGraw Hill, New York.
121. SCHMIDT-NIELSEN, K. (1965). *Desert Animals*. Oxford University Press, Oxford.
122. SCHOLANDER, P. F., WALTERS, V., HOCK, R. and IRVING, L. Body insulation of some arctic and tropical mammals and birds. *Biol. Bull.*, **99**, 225.
123. SCHULZE, R. (1963). Zum Strahlungsklima der Erde. *Arch. Met. Geophys. Bioklim.*, *B*, **12**, 185.
124. SELLERS, W. D. (1965). *Physical Climatology*. University of Chicago Press.
125. SIMMONS, R. C. (1970). *Heat and water vapour transfer of real and artificial leaves*. B.Sc. dissertation, University of Nottingham.
126. STANHILL, G. (1969). A simple instrument for the field measurement of turbulent diffusion flux. *Jl appl. Meteorol.*, **8**, 509.
127. STANHILL, G. (1970). Some results of helicopter measurements of albedo. *Sol. Energy*, **13**, 59.
128. STANHILL, G. and FUCHS, M. (1968). The climate of the cotton crop. *Agric. Meteorol.*, **5**, 183.
129. SUNDERLAND, R. A. (1968). *Experiments on momentum and heat transfer with artificial leaves*. B.Sc. dissertation, University of Nottingham.

130. SWINBANK, W. C. (1963). Long wave radiation from clear skies. *Q. Jl R. met. Soc.*, **89**, 339.

131. SZEICZ, G. (1970). *Spectral composition of solar radiation and its penetration in crop canopies.* Ph.D. thesis, University of Reading.

132. SZEICZ, G. and LONG, I. F. (1969). Surface resistances of crop canopies. *Wat. Resour. Res.*, **5**, 622.

133. TAGEEVA, S. V. and BRANDT, A. B. (1961). Optical properties of leaves. In *Progress in Photobiology*, ed. CRISTENSON, B. C. Elsevier, Amsterdam.

134. TAMM, E. and KRZYSCH, G. (1959). Beobachtungen des Wachstums faktors CO_2 in der Vegetationszone. *Z. Acker- u. PflBau*, **107**, 275.

135. TANI, N. (1963). The wind over the cultivated field. *Bull. natn. Inst. agric. Sci., Tokyo, A*, **10**.

136. TANNER, C. B. and PELTON, W. L. (1960). Potential evapotranspiration estimates by the approximate energy balance method of Penman. *J. geophys. Res.*, **65**, 3391.

137. THOM, A. S. (1968). The exchange of momentum, mass and heat between an artificial leaf and the airflow in a wind tunnel. *Q. Jl R. met. Soc.*, **94**, 44.

138. THOM, A. S. (1971). Momentum absorption by vegetation. *Q. Jl R. met. Soc.*, **97**, 414.

139. THOM, A. S. (1972). Momentum mass and heat exchange of vegetation. *Q. Jl R. met. Soc.*, **98**, 124.

140. TIBBALS, E. C., CARR, E. K., GATES, D. M. and KREITH, F. (1964). Radiation and convection in conifers. *Am. J. Bot.*, **51**, 529.

141. TOOMING, H. G. and GULYAEV, B. E. (1967). *Methods of measuring photosynthetically active radiation* (in Russian). Nauka, Moscow.

142. TUCKER, V. A. (1969). The energetics of bird flight. *Scient. Am.*, **220**, 70.

143. UCHIJIMA, Z. and WRIGHT, J. L. (1964). An experimental study of air flow in a corn plant air layer. *Bull. natn. Inst. agric. Sci., Tokyo, A*, **11**.

144. UCHIJIMA, Z., UDAGAWA, T., HORIE, T. and KOBAYASHI, K. (1967). CO_2 environment in a corn plant canopy. *J. agric. Met., Tokyo*, **23**, 1.

145. UCHIJIMA, Z., UDAGAWA, T., HORIE, T. and KOBAYASHI, K. (1968). Penetration of direct solar radiation into corn canopy. *J. agric. Met., Tokyo*, **24**, 141.

146. UCHIJIMA, Z., UDAGAWA, T., HORIE, T. and KOBAYASHI, K. (1970). Turbulent transfer coefficient and foliage exchange velocity within a corn canopy. *J. agric. Met., Tokyo*, **25**, 11.

147. UNDERWOOD, C. R. and WARD, E. J. (1966). The solar radiation area of man. *Ergonomics*, **9**, 155.

148. VAN BAVEL, C. H. M. (1966). Potential evaporation: the combination concept and its experimental verification. *Wat. Resour. Res.*, **2**, 455.

149. VAN EIMERN, J. (1964). Untersuchungen über das Klima in Pflanzengestanden, *Ber. dt. Wetterd., Offenbach*, No. **96**.

150. VAN WIJK, W. R. and DE VRIES, D. A. (1963). Periodic temperature variations. In *Physics of Plant Environment*, ed. VAN WIJK, W. R. North-Holland Publishing Co., Amsterdam.

151. VAN WIJK, W. R. and SCHOLTE UBING, D. W. (1963). Radiation. In *Physics of Plant Environment*, ed. VAN WIJK, W. R. North-Holland Publishing Co., Amsterdam.

152. VOGEL, S. (1970). Convective cooling at low airspeeds and the shapes of broad leaves. *J. exp. Bot.*, **21**, 91.

153. WAGGONER, P. E. and REIFSNYDER, W. E. (1968). Simulation of the temperature, humidity and evaporation profiles in a leaf canopy. *J. appl. Meteorol.*, **7**, 400.

154. WARREN WILSON, J. (1967). Stand structure and light penetration. III. *J. appl. Ecol.*, **4**, 159.
155. WEBB, E. K. (1965). Aerial Microclimate. In *Agricultural Meteorology*, ed. WAGGONER, P. E., American Meteorological Society, Boston.
156. WEBB, E. K. (1970). Profile relationships: the long-linear range and extension to strong stability. *Q. Jl R. met. Soc.*, **96**, 67.
157. WRIGHT, J. L. and BROWN, K. W. (1967). Comparison of momentum and energy balance methods of computing vertical transfer within a crop. *Agron. J.*, **59**, 427.

Index

Bold symbols indicate main entries.